D1488442

SMELTER SMOKE IN NORTH AMERICA

DEVELOPMENT OF WESTERN RESOURCES

The Development of Western Resources is an interdisciplinary series focusing on the use and misuse of resources in the American West. Written for a broad readership of humanists, social scientists, and resource specialists, the books in this series emphasize both historical and contemporary perspectives as they explore the interplay between resource exploitation and economic, social, and political experiences.

John G. Clark, University of Kansas, Founding Editor
Hal K. Rothman, University of Nevada, Las Vegas, Series Editor

SMELTER SMOKE IN NORTH AMERICA

The Politics of Transborder Pollution

John D. Wirth

 University Press of Kansas

© 2000 by the University Press of Kansas
All rights reserved

Published by the University Press of Kansas (Lawrence, Kansas 66049), which was
organized by the Kansas Board of Regents and is operated and funded by Emporia State
University, Fort Hays State University, Kansas State University, Pittsburg State University,
the University of Kansas, and Wichita State University.

This project is undertaken with the assistance of the Government of Canada.

Library of Congress Cataloging-in-Publication Data

Wirth, John D.
    Smelter smoke in North America : the politics of transborder
pollution / John D. Wirth.
        p.   cm. — (Development of western resources)
    Includes bibliographical references and index.
    ISBN 0-7006-0984-9 (cloth : alk. paper)
    1. Smelting—Environmental aspects—Canada. 2. Smelting—
Environmental aspects—Mexico. 3. Air—Pollution—North America.
4. Transboundary pollution—North America.   I. Title.   II. Series.
TD888.M4W57   2000
363.739'2—dc21   99-38225

British Library Cataloguing in Publication Data is available.

Printed in the United States of America
10 9 8 7 6 5 4 3 2 1

The paper used in this publication meets the minimum requirements of the American
National Standard for Permanence of Paper for Printed Library Materials Z39.48-1984.

In memory of my grandparents,
Gilbert and Hortense Davis
They followed the mines

CONTENTS

PREFACE

In part, this book traces back to a family history in Western mining. I was born in a Phelps Dodge company hospital in Dawson, New Mexico, because my grandfather, Gilbert C. Davis, was manager of the Stag Canyon Division that ran the Dawson coal mine. An efficient manager known for his contributions to mine safety, Mr. Davis was the boss. He could control everything in the town except the actual birth process itself, hence his enjoinder to my mother, Virginia, that his first grandchild be born not in Santa Fe, which was closer to the Los Alamos Ranch School where my father taught, but in Dawson. Later, as a boy of seven, I was impressed with the open pit mine and smelter at Morenci, Arizona, which my grandfather managed during World War II, and even more impressed when an army general presented him, his staff, and the men with an "E" for excellence in war production. A band played as this pennant was hoisted up the smelter's tall stack where it joined a navy "E" pennant already in place.

My brother Tim and I had the run of the corporate mansion in Clifton, formerly owned by the Arizona Copper Company, whose Scottish manager, Mr. Carmichael, left behind an enormous stuffed black horsehair chair that smelled funny and different. For years a signed photograph of P. G. Beckett, the vice president to whom Gilbert reported, was in his office beside a signed cartoon from the *Arizona Republic* depicting a fulminating John L. Lewis of the United Mine Workers, with whom it was my grandfather's unhappy duty to negotiate in his last assignment before retiring from Phelps Dodge.

Although I did not follow in my grandfather's footsteps, the pleasure of being where everyone knew Mr. Davis, including the small-town cafes and upstairs offices of old mining towns we visited, was a splendid preparation for the historian-to-be. I came to appreciate mining as a way of life, and to this day I harbor a respect for mining and the business and engineering challenges of process industries. Becoming an environmental historian in no way diminished the importance of this formation, even as the downside of mining's effect on the environment became apparent later on.

In part, this book originated out of my earlier work on Brazilian regionalism and my developing interest in North American integration and links with South America. The idea of doing a book on smelter pollution was prompted in the mid-1980s when I saw a map in *Business Week*

Wildcat strike during World War II. (Reg Manning cartoon, *Arizona Republic*, January 15, 1943; original endorsed by artist to Gilbert Davis)

showing the pollution pathway from a Mexican smelter moving north over New Mexico to the Colorado line and enveloping my favorite trout stream. This facility was Nacozari, a state-of-the-art copper smelter in Sonora that was starting up initially without controls. Meantime, the acid rain debate was running full force, and through contacts at the Environmental Defense Fund I became familiar with EDF's research into the effects of smelter pollution on high mountain lakes in the Rocky Mountains. This led me to Bisbee, Arizona, and to Dick Kamp, the multitalented founder of the Border Ecology Project that was leading the fight to clean up the Douglas smelter and two Mexican smelters known collectively as the Gray Triangle. In Dick's office was a much-xeroxed cartoon from the *Toronto Star,* depicting then prime minister Brian Mulroney looking put out with the news that his friend Ronald Reagan had just signed the first acid rain agreement, not with Canada but with Mexico. Sent by a Canadian nongovernmental organization to Bisbee, this cartoon triggered a North American epiphany: continental connections were being made.

In part also the book is informed by what I have learned as a founder, then president, of the North American Institute (NAMI) and by service on the Joint Public Advisory Committee of the NAFTA Commission for Environmental Cooperation. For the last ten years NAMI, a trinational public affairs group, has been examining the rapid emergence of North America as an integrated region. For its part, the commission was established to deal with the environmental consequences of that integration while pinpointing the many ways our three nations can address these issues through cooperative actions. As a participant, I have come to feel that, with respect to the environment, the internal borders of North America are now providing the social and economic context for a host of basically similar interactions even as their traditional function of defining relations between states declines. From this basic insight arose the recently published *Environmental Management on North America's Borders,* for which I was cocontributing editor with Richard Kiy, and I have incorporated Canada into my teaching and writing on Western Hemisphere integration as well. Thus it is no accident that the Trail smelter in British Columbia, which gave rise to the first major case of transborder air pollution in North America, takes pride of place in this book.

The result of all these strands and influences coming together is the first book I know of to explore in depth an aspect of the environmental history of North America within the context of a new continental space. Indeed, globalization makes such an approach inevitable. Yet developing a conceptual vocabulary to think about cross-border regionalism is still quite novel. While researching this book I soon came to appreciate that the mining and smelter industry had been thinking continentally since the 1890s, almost a century ahead of the institutions to deal with the

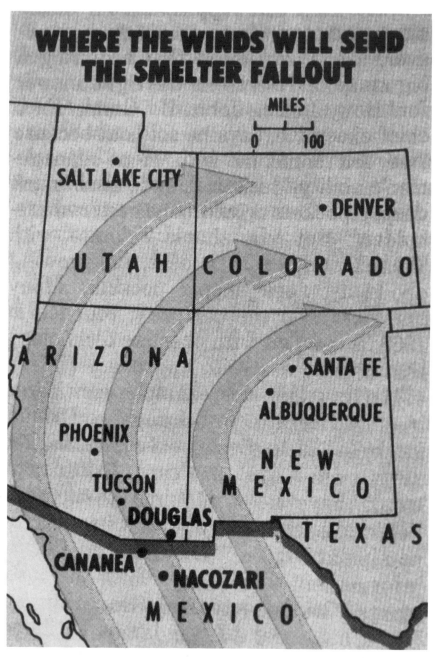

Smelter pollution from the border. (Reprinted from July 22, 1985, issue of *Business Week* by special permission, copyright © 1985 by The McGraw-Hill Companies, Inc.)

Prime Minister Brian Mulroney of Canada on learning that the United States has signed first with Mexico. (John Larter cartoon, *Toronto Star*, January 8, 1987. Reprinted with permission—The Toronto Star Syndicate)

transborder effects of industrial pollution. How it was that the two realms of business and public environmental policy intersected and ultimately came together in North America is the subject of this book.

History matters, and to recuperate the world of mining and smelting where borders were often brushed aside and engineers and managers shared a common view of North America is one of my purposes in writing this book. To be sure, the literature is vast with respect to mining itself and the labor disputes that arose in this industry. Readers interested in these aspects should follow the well-marked paths into these subject areas, although I do touch on mining and labor history when appropriate. When it comes to the environment, however, we enter new territory. Here the literature is sparse, and few have yet explored the archival sources in all three countries on which continental history in general, and continental environmental history in particular, must be based. In this respect, the book is a pioneering work.

Thematically, there is a point-counterpoint at play. I refer to the slow but ultimately successful development of public policy to contain and control smelter pollution in the face of industry's determined efforts to maintain production even if this meant (as it did) maintaining the right to pollute. Here again, there is a vast legal and scientific literature bearing on this subject, but much of it is useful primarily as the raw material for the environmental histories that are for the most part still to be written. The archives themselves are largely untapped. In fact, from the way historians look at change and causality, much that is useful has not yet been said in the detail and specificity that can only come from doing the archival research into this interplay of public policy and corporate decision-making in the matter of industrial pollution.

For example, the famous Trail smelter case lives on in the legal literature as one of the fundamental building blocks in the law of international injury. I refer to the "polluter pays" principle that is always associated with Trail. But in fact *this principle was not the main significance of Trail*, as I show in the pages that follow. When I discussed this finding with a corporate friend with a strong environmental ethic, he expressed annoyance that I would tinker with the underpinnings of such a useful legal concept. A colleague at Stanford Law School liked the work but remarked, "Once again things are not what they seem." These reactions in different ways provided insights into the significance of this history, let alone even how the book may be received. In fact, the approach is sufficiently novel that it would be useful to summarize the main argument, as follows:

The Trail smelter dispute—a precedent-setting case in international law on transboundary air pollution—was not simply a diplomatic spat between the United States and Canada, as it is often depicted. To begin with, smelter industries on both sides of the border shared a common interest in adjudicating the case in a manner that would allow them to continue polluting the atmosphere with minimal regulatory restrictions. They therefore cooperated with each other in the accumulation of scientific data that supported the political goals they hoped to achieve.

Furthermore, Canadian government scientists were yoked together with industry, while U.S. scientists did not speak with one voice. Instead, they reflected divergent agricultural and industrial interests. Department of Agriculture researchers focused on the effects of smelter smoke on vegetation—trees, plants, crops—in search of *visible* (burned leaves, the effects of declining soil productivity) and *invisible* (stunted growth, lower food value in the case of grains and grasses, and so forth) damage. They sought data that would allow them to comprehensively assess how smokestacks affected crops and farmland. Bureau of Mines researchers, by contrast, tended to sympathize with the smelter industry. They focused on finding "techno-fixes" that would reduce, but not eliminate, the side effects of smelting without compromising productivity. Contextually, this science was reified by a legal paradigm that placed proof of damage at the very center of the exercise.

Moreover, the Trail smelter was not the only smelter on a U.S. border, a fact about which industries and diplomats in both countries were painfully aware. If the United States won against Canada at Trail, air pollution coming from Detroit industries into Ontario could be held accountable as was pointed out in an adroit use of linkage politics by Canadian negotiators. And, if U.S. firms were found liable for damages they were causing on Canadian soil from Midwest industries, smelting interests would be liable for similar damages on Mexican soil south of the Arizona border.

All of these factors tipped the balance in favor of a narrow political and legal resolution of the Trail case: U.S. farmers downwind from Trail received some economic compensation for crop loss, and the Trail smelter came under a mild regulatory regime, but no thorough research on invisible damage was continued, and no fundamental principles were laid down for eliminating transboundary pollution.

The Trail precedent, in turn, made it easy for Phelps Dodge, with its Arizona border smelter at Douglas, and the American Smelting and Refining Company (ASARCO), with its Texas copper and lead plant in El Paso, to continue polluting Mexico for decades thereafter while paying small compensation fees, in the case of Douglas, to affected Mexican and U.S. farmers. It would take a whole new effort—coming in large part from the environmental movement of the 1960s—to outflank the Trail precedent and bring the smelter industries under a more stringent regulatory regime based on new scientific findings into the effects of smelter pollution on human health as well as plants, and on the newly discovered phenomenon of acid rain.

Protecting human health from the effects of industrial air pollution was front-loaded in the Clean Air Act of 1970, while assessing damage to property became less important. This approach enabled government officials and nongovernmental organizations (NGOs) to outflank the corporate lawyers, whose tactic of choice was reduced to narrowing, when they could not overturn, the criteria in endless litigation. Furthermore, the Environmental Protection Agency was created in 1971, opening space for NGOs to advance the environment. This combination of institutional and social power changed the balance of forces, as is abundantly clear in the case of the Douglas smelter. And for its part Mexico found good reasons for cooperating with the United States in a cross-border agreement to control smelters in Sonora as well as Arizona. Signed in 1983, the La Paz agreement antedated NAFTA by a decade.

Finally, the two cases are significant not only as history but also for policy formation today. Thanks to the experience gained by all three countries in dealing with smelter pollution, developing a truly North American institutional response to continental air pollution is no longer a distant goal. As of current writing, the three nations are reaching agreement on criteria to monitor and measure airborne pollutant pathways; they may soon sign a Transboundary Environmental Impact Assessment agreement that should abate future cross-border pollution from new plants before it starts; and they are gaining experience with interagency cooperation at all levels of government. Air pollution specialists in Canada, the United States, and Mexico communicate and share information as readily today as their mine and smelter counterparts in industry have been doing for a century. Industrial practices and environmental policies are now yoked together. And while it may be a stretch to say that this extractive industry has become "environmentally friendly," experience gained in the last thirty years bears out the maxim that good environmental management is good management.

I wish to thank the Stanford undergraduate program in Washington for providing me with the opportunity to research the National Archives' holdings on the Trail smelter while I was in residence giving a seminar on the origins of NAFTA in the fall of 1993. The same goes for Stanford in Santiago, where I did the final editing on this book while presenting a course on NAFTA and Mercosur. The Canadian National Archives helped in many ways with the Trail research and is a superb place to work. The British Columbia Archives in Victoria was also accommodating. To COMINCO in Vancouver and Phelps Dodge in Phoenix I gratefully acknowledge access to corporate records, and to Senator Jack Austin and to Bob W. Loughlin I am grateful for facilitating this access. Records at EPA's Region 9 in San Francisco were made available to me by Dave Howekamp, whose role in the Arizona-Sonora case is chronicled below. Alejandra La-Jous Vargas and her father, Adrián LaJous, facilitated my research in Mexico in many ways. Stanford's University Archives and the Hoover Institute provided materials on the life and work of Robert Swain, a pioneer in the science of air pollution who brought early recognition to the chemistry department. In Tucson, the University of Arizona gave me access to the newly cataloged archive of Southwest Environmental Service, an NGO whose founder, Priscilla Robinson, became both a mentor and a friend. Dick Kamp and the EDF's Bob Yuhnke were both generous with their time, documents. and insights. The other people I interviewed in connection with this research, listed in the bibliography, all gave me essential parts of the story; thanks to each of them as well. Comments from the two readers who read the manuscript for the University Press of Kansas helped to sharpen the argument, a much needed and welcomed contribution on their part.

I wish to thank my colleagues on the Joint Public Advisory Committee of the NAFTA Commission for Environmental Cooperation in Montreal for the privilege of exploring together new ways of thinking and new policies and practices to advance environmental cooperation among the three nations of North America. And for the many insights I have gained into the heart and soul of the developing North American community in the meetings of the North American Institute over the last ten years I am especially grateful to founding president Susan Herter, Senator Jack Austin, Jesús Silva Herzog, Robert L. Earle, Richard Kiy, and Sanford Gaines.

This book is dedicated to my maternal grandparents, who began their married life in the dusty coal towns of Colorado, Wyoming, and New Mexico and, while climbing the Phelps Dodge corporate ladder, did their part to enhance American productivity and win the Second World War. Grandfather Gilbert would know that this book is not antimining. Rather, times change. Grandmother Hortense kept house in these remote and bleak surroundings. She planted roses everywhere they went, including at the house in Clifton, Arizona, with its spectacular rose garden. She, too, would understand my motives. Concerning the environment, we learn from the past, which tells us we can and must develop a truly North American institutional response to the challenge of air pollution to the continental commons.

ABBREVIATIONS

ADHS	Arizona Department of Health Services
ASARCO	American Smelting and Refining Company
BACT	Best available control technology
BC	British Columbia
CEC	Commission for Environmental Cooperation
CODELCO	Corporacíon Nacional del Cobre (Chile)
COMINCO	Consolidated Mining and Smelting Company
CPA	Citizens' Protective Association (Northport)
CPR	Canadian Pacific Railroad
EDF	Environmental Defense Fund
DRW	Douglas Reduction Works
ENGO	Environmental nongovernmental organization
EPA	Environmental Protection Agency
GASP	Group Against Smelter Pollution
IADB	Inter-American Development Bank
IBWC	International Boundary and Water Commission
IJC	International Joint Commission
INCO	International Nickel Company
MERCOSUR	Common Market of the South
MPR	Multi-point rollback
NAAQS	National ambient air quality standards
NAFIN	Nacional Financiera (National Development Bank)
NAFTA	North American Free Trade Agreement
NCAC	National Clean Air Coalition
NGO	Nongovernmental organization
NRC	National Research Council (Canada)
NSO	Nonferrous smelter order
PD	Phelps Dodge Corporation
PPM	Parts per million
SES	Southwest Environmental Service
SCS	Supplemental Control System
SEDUE	Secretaría de Desarollo Urbana y Ecología (Secretariat of Urban Development and Ecology)
SIP	State implementation plan
SRE	Secretaría de Relaciones Exteriores (Secretariat of Foreign Relations)

SX-EW   Solvent extraction and electrowinning leach process
TEIA   Transboundary environmental impact assessment
USBM   U.S. Bureau of Mines
USDA   U.S. Department of Agriculture

PART I: THE TRAIL SMELTER
DISPUTE, 1927–1941

Smoke drifting over the international boundary from the lead and zinc smelter at Trail, British Columbia, caused damage to crops and forests in the state of Washington. From the start, in 1925, Consolidated Mining and Smelter Company accepted responsibility for this damage and was willing to compensate the landowners and farmers in Stevens County while installing advanced fume-control technologies to lower the daily tonnage of sulfur dioxide emitted from the plant, thus limiting damage. The farmers, in turn, formed a Citizens' Protective Association and demanded higher payments by virtue of collective action than they otherwise could have received individually; they also demanded that the smoke "visitations" cease. At issue was the amount the Montreal-based company was prepared to pay short of making the plant unprofitable to operate or closing. To this extent, the dispute was very similar to earlier confrontations between U.S. smelters at Anaconda in Montana, Ducktown, Tennessee, and Salt Lake in Utah and the agricultural interests damaged by industrial pollution.[1]

Yet from entirely conventional beginnings, this industrial dispute quickly escalated to the highest levels of the Canadian and U.S. governments. In short order, it raised an unusual and difficult set of policy issues. When the Trail smelter dispute was finally settled in 1941, fundamental issues had been raised in such areas as international law, the technology of air pollution abatement, and the scientific investigation of the effects of sulfur dioxide (SO_2) on forests, crops, and soils. Today the Trail settlement is justly celebrated as the first international ruling on transborder air pollution and for its affirmation of the "polluter pays" principle in international law. However, the aspects of the case touching on business decisions, pollution abatement technology, and the science have largely been forgotten. That the farmers and landowners of Stevens County received only a fraction of what they claimed for damages is also

1

forgotten to all but their descendants. At heart, this case turned on the assessment of liability for damage caused by air pollution. Once settled, a number of other issues raised in this complex and difficult case were dropped or faded from the literature. Moreover, the significance of the case as a North American, indeed a *continental* issue—which was very well understood by government and industry at the time—has been lost entirely. In probing the reasons for this selective memory, it becomes clear that the Trail smelter dispute is both more and less than meets the eye. And while the primary source material is extensive, what follows is the first historical analysis of the Trail dispute based on the full range of documents in Canadian and American archives.[2]

In broad overview, the following dynamics were at play. As mentioned, the dispute was similar in origins to others in the United States, but two factors stood out from the start. Trail was the largest smelting complex of its kind in the British Empire, and the visibility and power of Canadian Pacific Railroad, the parent corporation, assured that the interests of Consolidated Mining and Smelter Company (C.M.&S., later COMINCO) would be considered national interests by Ottawa. And while the smoke drifting down the Columbia River Valley knew no borders, it crossed the international frontier just fourteen miles south of the works and the nearby town of Trail, which internationalized what would otherwise have been a local or at most a subregional issue.

Furthermore, the area is heavily mineralized, and the smelter claimed with reason that there would be no agriculture were it not for the markets created by the plant. Therefore, industry must take pride of place. Landowners and farmers—referred to as "settlers" in the documents—saw things differently. Since the 1890s, they had been expecting to benefit *both* from rising land values and increased mining and smelting activity. But shortly after World War I, the smaller lead smelter at Northport shut down, while the much larger lead and zinc works at Trail expanded production. Stacks were raised to four hundred feet in an effort to disperse the increased pollution load. Both parties were disagreeably surprised when this approach did not work: prevailing air currents caused a marked increase in diurnal downdrafts, scorching crops, accelerating forest loss, and filling the Columbia River Valley below the plant with choking, noxious fumes. The taste and smell and sight of polluted air became part of peoples' daily lives. The stage was set for confrontation.

Thus from the start Consolidated management was fighting more than a classic smoke action by a few score small producers, or "smoke farmers," who from the company's perspective stood to gain more from suing the largest industry in British Columbia than they ever could make selling timber and produce from their marginal farms along the Columbia River. Having settled similar damage claims on the Canadian side of the line, Consolidated would have preferred to buy off the U.S. interests with cash payments, by entering into smoke easements on the properties affected, or by outright purchase. However, the latter two of these standard defensive measures were barred to it because Washington's state constitution prohibited foreigners from owning property. And while some farmers accepted payments, the majority, their bargaining power enhanced by the company's inability to purchase land, stuck together and by 1926 were petitioning their state and federal representatives for support against a giant foreign corporation whose "fumigations" were damaging U.S. lands.[3]

Congressmen and senators asked the State Department for guidance and pressed for action to redress wrongs to their constituents. Because this was a boundary problem, raising issues of extraterritoriality, the state and provincial governments were loath to get deeply involved. In the celebrated Tennessee Copper dispute of 1912 (Ducktown), where transboundary pollution of forests in the state of Georgia was at issue, the U.S. Supreme Court ruled that the polluter across state lines must pay. But Trail was no domestic issue. In short order, the two federal governments became parties to the case, first through the International Joint Commission (IJC) and then through an Arbitral Commission that sat from 1937 to 1941 until the dispute was finally resolved.

Thus what began as an entirely conventional dispute between two local interests, the smelter pitted against agriculturalists, was quickly internationalized, which greatly enhanced the importance of the dispute, both by widening the range of players and also by expanding the potential impact of the case itself. For one thing, the settlers had a powerful new ally on their side, the U.S. State Department, which in 1928 engaged a team of scientists from the U.S. Department of Agriculture to conduct the scientific investigations. For another, Consolidated, despite its stable of in-house experts, felt in danger of being outgunned and turned to Canada's National Research Council (NRC) for scientific support. In addition, Trail's owners had access to the Salt Lake research station, maintained by American Smelting and Refining Company for the industry as a whole, and to the network of university scientists receiving grants from industry. Thus, the fact that Trail quickly became a foreign policy issue between two sovereign states pitted against each other in a legal dispute brought new actors to the scene and changed the role of others.

The two governments agreed to arbitration, turning first to the IJC, which since its founding in 1909 had limited its purview to boundary and water issues. However, as the one existing transboundary institution the IJC was asked by both governments to investigate and then recommend a settlement to this industrial dispute. Canada itself had just emerged from the shadow of the British Foreign Office; Trail was the first diplomatic

dispute it was handling entirely on its own.[4] In 1931, the IJC awarded a
$350,000 damage settlement to the farmers, far less than they had been
asking for and far more than the company thought they deserved. But
there was no ruling that the transport of sulfur dioxide causing damage to
American territory must cease entirely, a key U.S. demand, and the farmers pressed for a higher damage award. Under pressure from Washington
State's congressional delegation, the State Department took the highly unusual step of rejecting the IJC settlement. A three-person Arbitral Tribunal
was established in 1935 to resolve the by now long-running dispute.

In 1938, the tribunal ruled narrowly on the legal merits of the case,
granting the farmers only $78,000 in additional damages to the IJC award
while rejecting or setting aside as unresolved most of the scientific case
presented by the United States (see Chapter 2). At the same time, the tribunal placed the smelter on an operating regime that in addition to fume
abatement at the stack and coupled with acid recovery measures operating at the plant substantially reduced the cross-border pollution. With a
war on by then, both sides were eager to get on with the production of
vital lead and zinc supplies to the British government, and the Trail dispute had already faded from the public policy arena when the case was
closed in 1941.

The upshot is that the farmers eventually received only a fraction of
their claim for damages, while the plant was placed under a tight, but by
no means punitive, operating regime to control for sudden downdrafts of
early morning fumigations during the growing season. In the smelting industry, this dispersion technique came to be called a Supplemental Control System (SCS). The SCS combined continuous ambient SO_2 monitoring
and meteorological forecasts to reduce production when ambient SO_2 concentrations reach elevated levels. Usually, but not always, it was considered a supplement to acid recovery measures at the plant and tall stacks.
In the course of establishing this regime, much was learned about the meteorological aspects of pollution transport. Trail management had sought
to avoid the imposition of any operating regime other than its own. However, the tribunal-imposed regime was far better than having to accept an
absolute limit on SO_2 emissions that the United States had wanted to impose. In the Trail case, U.S. and Canadian smelters also sought to prevent
the imposition of an industry-wide standard, which they also achieved.

The IJC established a large award but no operating regime because
Consolidated had convinced the commission that it had the technology
to substantially reduce crop and forest damage caused by SO_2 emissions.
In fact, from the start of its dispute with the Stevens County farmers in
1926, the company's research department was seeking ways to reduce
stack pollution while accommodating increased production. Acid plants
were installed; the production of artificial fertilizers and elemental sulfur

Photo of the Trail smelter in 1923, showing the heavily wooded countryside and the smelter plume in the Columbia River Valley. (British Columbia Archives)

found markets; and a stack dispersion regime was put in place. By 1937, the release of SO_2 into the atmosphere had been cut by two-thirds, from 1,000 tons a day to 350 tons a day. The tribunal was impressed but wanted even more control over emissions to accommodate American demands. The tribunal-imposed dispersion regime handled most of the rest, greatly limiting crop and forest damage even though the plant was operated at full capacity for war production. (Concerns about the long-range transport of sulfates and nitrates, the building blocks of acid rain, arose only in the 1970s.) Moreover, the chemical and fertilizer plants that converted sulphur and other by-products of the smelting process to commercial use were demonstrably successful and profitable.

In this way the company argued successfully against a second large damage award and fought off U.S. demands that all pollution cease. Instead, the problem was essentially solved by technical measures: for example, chemical recovery using refinements on existing technology and meteorological measures with emphasis on the operating or SCS regime. This solution was enough to satisfy the tribunal and was far more than any other smelter in North America had done to control point-source emissions. Consolidated emerged with its reputation for progressive management enhanced. It was a highly profitable producer into the 1960s.

Yet in a real sense Trail, and by extension the North American smelting industry as a whole, won too big a victory, while society received too little compensation in terms of a deeper awareness of the dangers of air pollution to health and the environment. Scientific investigation into the

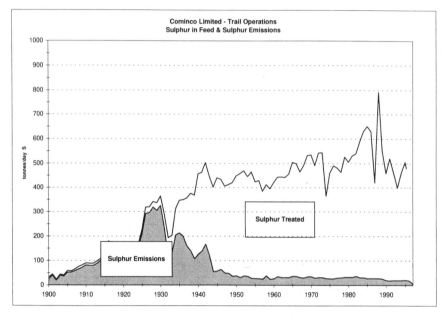

Sulfur recovery at the Trail smelter. (Courtesy of COMINCO)

damage caused to trees, plants, and soil by what came to be called "dry" acid deposition slowed and did not resume in earnest for a generation. The gap between what scientists could already say about the effects of pollution and the very narrowly defined legal remedies given standing in the courts to abate pollution was still huge. In short, there was a disconnection between science and the law.

Once resolved, Trail resulted in the decline of scientific interest in the complex chemistry of point-source pollution from metal smelters. Given this result, would it be too much to call this aspect of the case a failure? It fell to the next generation to clamp down on the residual pollution that dispersion regimes did not eliminate and to grapple with what by the 1970s was thought to be the much bigger and more serious problem of acid rain, or long-range acid deposition.[5] By then, the continental dimensions of the problem were truly inescapable as well.

It can be argued that the Trail settlement, which was in fact narrowly limited to the single case at hand, actually set back the development of public policy measures to deal with industrial air pollution. While the dispute was active, Trail created a climate favorable for the development of policies to ameliorate pollution at industrial sites in North America. However, the lessons learned about the human and environmental costs of air pollution were small or negligible. The science-driven Clean Air Act

The Trail smelter in 1930, with smoke going down the Columbia River Valley. (U.S. National Archives)

of 1970, with its central mandate to protect human health, was a generation away, and the Trail case did not prefigure it or its 1990 successor that set even higher environmental standards. Furthermore, for all of the money spent on scientific investigations and the hearings themselves, the actual results in terms of both precedent for damage recovery and the legal standard for pollution abatement were both impressively narrow.[6]

To unravel the meanings of this complex case, I have organized the narrative into three chapters. In the first chapter the gathering dispute between C.M.&S. and the farmers is placed in the context of the specific geographical setting and transboundary culture of the upper Columbia River Valley, followed by analyses of legal aspects of the case and concluding with an assessment of technology and the business aspects of the dispute. The second chapter is an in-depth evaluation of the scientific evidence, perhaps the most troubling aspect of the dispute. Finally, in the third chapter I examine the continental strategies and perspectives of the nonferrous smelting industry in North America.

To revisit Trail, then, is to return to an earlier era of limited environmental consciousness, a time when to most people the smell of smoke was still the smell of money and jobs. By today's standards, the absence of nongovernmental organizations or the science-driven regulatory regimes that have operated in Canada and the United States since the 1970s is striking. However, serious questions about the environmental and human costs of smelters were in the air even then. If protecting nature was still far from a moral mandate, it was nonetheless acknowledged that polluting industries should bear a cost for, and not just profit from, the harm they caused. Just what that cost should be is the essence of the case narrated below.

CHAPTER ONE

Settlers and Smelters in the Columbia River Valley

The once remote and heavily forested Kootenays were opened to settlement when the Spokane Falls and Northern Railway brought homesteaders and development to the upper Columbia River and its tributaries flowing south from British Columbia into American territory. After the line reached Northport, Washington, in 1892, it branched out to the mines of Rossland and Nelson, in British Columbia, where it connected with the Canadian Pacific Railway (CPR). Mining, timbering, and small-scale agriculture grew up to serve the American-owned copper smelters at Northport and Trail, British Columbia, both founded in the 1890s. Moreover, most of the lead and zinc ores from this heavily mineralized region were sent south to Spokane for processing.

The economic futures of the two smelter towns soon diverged. Operated by the Le Roi Mining and Smelting Company, the Northport smelter ran intermittently from 1898 to 1908; later, under the Northport Smelting and Refining Company, it was a lead smelter during World War I. Unable to secure long-term ore contracts, it shut definitively in 1921 with the loss of three hundred jobs. Meanwhile, the Trail smelter (owned by the CPR since 1898) grew into the world's largest lead and zinc complex, with a payroll of some four thousand jobs and a large market for local farmers. Spokane lost its preeminence. Instead of being the hinterland to American-owned processing plants—as happened to the Sonora mines on the Arizona border—the southeastern corner of British Columbia became the growth pole for the mining and smelting industry in that region.

In the boom-and-bust cycles so typical of the West, Canadians and Americans did relatively well in this remote corner of British Columbia, but settlers on the American side of the line were disappointed. Their hopes for rapid economic growth centered on the Northport area, for its industrial potential and because the area seemed well suited to agriculture, including fruit orchards. The rapid growth of Trail was a disappointment to them, but hopes for a speculative boom based on mines and land development were sustained through World War I.

Indians were the first big losers. Lakes Indians migrating across the line from Canada had adopted to farming on the Columbia River benchlands after the Hudson's Bay Company abandoned Fort Coleville. In 1896, the Colville Indian Reservation across the river from Northport on the west bank and running to the international boundary was opened to gold prospecting. As settlers encroached, the natives tried and failed to block ferry service access from Northport. By 1900, the government had opened large areas of the reservation for settlement, and a land rush ensued. Homesteaders filed on timber claims, and logging and lumbering began in earnest.[1] Small farmers moved onto the benchlands on either side of the Columbia River canyon. In 1910, some fifteen thousand acres of former Indian lands were purchased by the Upper Columbia Land and Fruit Company, a New York corporation, which also installed extensive apple orchards downstream at the town of Marble. Speculators hoped to replicate the fruit boom in Wanatchee and the Yakima Valley.

As a classic resource frontier promoted by the railroad, northern Stevens County had been expected to boom with rising real estate values and a growing population. Northport would be the growth pole. However, by World War I it was clear that the area would not become a major fruit-producing region: the Upper Columbia Land and Fruit Company's orchard at Marble failed, and large-scale colonization of its lands near Northport never occurred. Lumbering was the most important industry, but the best and most accessible stands of yellow pine, Douglas fir, larch, and cedar were soon logged out, and fires destroyed much other timberland all the way to Marble. By 1929, only three small sawmills survived of the eighteen mills that had been operating in the area. Northport's population rose and fell with the fortunes of the resource economy, from the 787 recorded in the 1900 census, declining to 476 in 1910 with the smelter idle, and up to a high of 906 in 1920 when the smelter was again in operation.[2]

The final closing of the Northport smelter and the loss of payroll coincided with a major agricultural depression in 1921. "In the years intervening between 1922 and 1929," a USDA researcher reported, "a number of sawmills and mines ceased operations which threw men out of work and cut off a source of revenue on which many farmers had been dependent. With opportunity for work and income gone or reduced, many persons left the area, properties were abandoned and taxes not paid."

The mid-1920s were also drought years, which stressed the remaining forest stands, adding to the risk of fire and beetle damage while lowering farm output. Then the Great Depression of 1929–1937 added to this litany of economic woes, intensifying the severity of the first depression and bringing "calamity to the area through falling prices of natural products, loss of markets and a further drop in employment, all of which mean[t] reduced purchasing power in the area." By 1930, Northport's population was down to 391, a mere two-fifths of what it had been ten years before.[3] Almost one-quarter of the thirty-three thousand acres under cultivation a decade before had been abandoned.

Those who stuck it out in the rural areas surrounding Northport were mostly middle-aged or elderly people, as the country relief supervisor reported in 1934:

> The relief load is largely made up of rural people, who are living on sub-marginal lands, and who only make a decent living in the best of times. . . . These sub-marginal residents or "economically stranded" people live up in the foothills or in the small valleys, the so-called "stump ranchers." Frankly, most of these people are those who are least able to cope with the economic struggle. . . . I would estimate 125 families in the county who are so "economically stranded."[4]

Some farmers scraped a meager living from their small farms on the benches bordering the river, while others, perhaps the worst off, had already reverted to subsistence farming after the smelter closed and lumbering declined, earning their living by getting out a few ties, poles, or by doing government work on the roads.[5]

Meanwhile, bitterness was felt at the success of Trail, which was blamed for the Northport smelter's decline. But in mining territory hope springs eternal, and the expectation was that another plant would come in soon, providing jobs and a market for the farmers. The Northport district was "frequently considered or referred to as being potentially the richest mineral region in the American Northwest. So far no mining/smelting plan has developed for Northport," wrote Stewart W. Griffin, a USDA scientist, in 1928, "but everyone is still waiting for it." Boom talk died down with the Depression.

However, in 1935 the old lead smelter was sold to the Aurum Mining Company, and it was rumored that a gold refining plant or a cement factory would begin operating in Northport. There were rumors that the Consolidated Mining and Smelting Company, the CPR subsidiary that operated the Trail works, was preparing to move in. Consolidated already owned the Northport Power and Light Company, whose power sales had declined drastically after the Northport smelter shut down. The new Grand Coulee Dam, with its power generating capacity, and the filling of Lake Roosevelt behind it almost all the way upriver to Northport might attract a smelter. "Why not put this in Northport, an ideal location for a smelter?" wrote Ed Morris, owner of the Northport drugstore and a local booster. The area was "already ruined for agriculture . . . [from pollution and] the new smelter wouldn't have to pay for crop and timber damages" from smelter smoke, he advised Washington.[6]

Meantime, the Colville Indians lost heavily again, as the rising waters of Lake Roosevelt flooded more than twenty-one thousand acres of prime bottomland, some of it on the reservation. Grand Coulee Dam blocked salmon access to more than one thousand miles of productive river. Kettle Falls, a premier salmon run and the spiritual centering of Indian culture, disappeared under the rising waters of Lake Roosevelt and with it some of the Indians' best hunting, farming, and root-gathering land. Tribal burial grounds went underwater. "The tribe had to move their school at their own expense, lost telephone service along the shore that was not restored until 1975, were stuck with some of the highest electricity rates in the state and got no compensation for their lost salmon." Indian unemployment hovered around 50 percent for decades. Finally, in 1994, the federal government announced that it was settling the long-standing Colville claim with a $54 million lump-sum payment for damage caused by the dam, plus $15.3 million a year in annual compensation for the lost fishery and lands.[7]

It was in this climate of rumors, economic decline, and the emptying out of a once promising frontier region, involving at most a few hundred people, that the celebrated Trail smelter case arose in the mid-1920s. This transborder environmental dispute engaged officials at the highest levels of both governments, energized no less than four teams of government and company scientists, and engendered the first international ruling on transboundary air pollution. The causes go back to Trail itself, to the conditions prevailing on the other side of the border in British Columbia.

Purchased for something less than $1 million by the Canadian Pacific Railway in 1898, the small Trail smelter was expanded by Walter Hull Aldridge, an American mining engineer who headed CPR's mining and metallurgical interests. Aldridge developed the business plan and the management team that transformed Trail into a premier producer of lead and zinc. In 1905 the smelter and its mines were incorporated as the Consolidated Mining and Smelting Company, with CPR controlling half the stock. That the Canadian Pacific financed the company's growth was a big advantage from the standpoint of funding and developing an integrated operation. CPR controlled the rail link to the Sullivan mine at Kimberly, which was acquired from the American Smelting and Refining Company in 1910, thus locking up very large deposits of lead and zinc ores (which produced well into the 1960s) and breaking the hold of American lead smelters in Spokane over the processing of Canadian ores.

This move to Canadianize lead production was bolstered by tariff protection. Coking coal, and phosphate deposits as well, came from the nearby Crow's Nest Pass. Ore suppliers' contracts were diversified, including some with U.S. mines. Furthermore, rather than deadhead the ore cars back from Montreal, the CPR wanted to ship agricultural products to British Columbia. As part of its nation-building agenda, the railroad became a large purchaser of flour, meat, and dairy products from the prairies and in so doing developed markets in the wheat fields for smelter by-products such as copper sulfate and later for the superphosphate fertilizer that it sold under the famous Elephant brand. When Aldridge left in 1910, C.M.&S. was the largest industrial employer in British Columbia.[8]

By 1930, Trail was a flourishing community of twelve thousand people, international in its workforce but intensely local in its preoccupation with small-town events, and with a strong sense of place that marked it instantly as a typical mine and smelter town. Revisiting Trail in 1943, Stanford professor Robert Swain found that the

> town has not changed appreciably. The fishermen line up as usual each evening along the river parapet; the cheerful old paper man who asks you to "buy a *Vancouver Sun* and get to be a well-informed man," is still there; and in the Crown Point Hotel restaurant there is the Chinese boss, two of the same waitresses, the same mimeographed menu whose every item one could recite from memory; and actually the same prices of old.[9]

It was and is a company town, remembered by many as a good place to grow up despite, and because of, the ever-present plume.

Under the forceful leadership of its new superintendent, Selwyn Gwillym Blaylock, familiarly known as Blay, the company expanded operations while becoming an industry leader in the application and development of new technologies to process the Sullivan ores, which were high in sulfates. The American Institute of Mining and Metallurgical Engineers awarded Blaylock its Douglas medal for his contributions to metallurgy. Blaylock ran a one-man show. Few aspects of this community escaped his interest, whether it was Trail's champion hockey team—the Trail Smoke Eaters, who played against the Kimberly Dynamiters and Spokane—or the in-house company union, which despite his determined efforts finally lost control to the C.I.O. in 1940.[10] This smelter was "no candy factory," he reminded labor organizers. (In Arizona James and Walter Douglas, Montrealers who were mainstays at Phelps Dodge Corporation, were paternalists of the same ilk.) And when local Canadian farmers brought suit against his company for damaging crops and timber, Blaylock mounted a forceful and determined defense.

Smelters at Trail and Northport shared the same forested region, initially an advantage because both plants began operations with "heap roasting," an open-air reduction process that required vast amounts of wood fuel while heavily damaging these same forests with SO_2. To control the smoke, stacks were built at both plants, and Trail also installed Cottrell converters that used an electrostatic process to capture flume dust, particularly lead particles that were a recognized health menace to the workforce. The Northport smelter was smaller and less technologically advanced. From the start, American farmers and timber interests sued it for damages, and the smelter paid out damage awards and purchased "smoke easements" from the settlers, as a standard cost for doing business, until it shut in 1908. As well, Trail handled its own litigation problems by buying up lands within five miles of the plant and purchased smoke easements from local farmers that removed the threat of litigation when a burn occurred.

Reopened as a lead smelter, the Northport smelter was sued again in the state superior court, but this time it fought off damage awards thanks to Lon Johnson, a prominent local attorney who was aided in his defense by a stable of smelter experts provided by the North American smelter industry. By 1916, Trail was also in litigation and used basically the same defense appearing in the Northport court records, namely, that the damage to crops and forests was likely due to natural conditions and poor farming methods rather than smelter fumes, but if SO_2 damage to agriculture and forests could be proved, the company would pay.

To evaluate the damage, if any, to these Canadian farmers, C.M.&S. hired Dr. P. J. O'Gara, a pollution expert working for the American Smelting and Refining Company at their research station in Salt Lake City. R. C. Crowe, the company's attorney, fought one suit all the way to the Canadian Supreme Court before settlement. With its expanding operation, Trail was a big target. Men claiming expertise about the effects of SO_2 on crops, animals, and human health offered their services to the Canadian farmers, some fifty-two of whom began a "smoke action." The suits went to arbitration by the provincial court, and Consolidated paid some $60,000 to buy the farmers out or to neutralize them by purchasing smoke easements. From the company's point of view, being bled by "smoke farmers"—men who had more to gain from extorting payments than by selling produce to the very market created by the smelter—was always a danger. From the farmers' point of view, there was little option but to settle. By the early 1920s, Consolidated owned most of the valley floor down to Columbia Gardens, near the U.S. border. Champion cattle were raised on the abandoned farmland, and a model farm was established to show that farming could be conducted profitably in the lea of a large smelting complex.[11]

Interestingly enough, farmers on the American side of the line testified for the company, affirming in provincial court that their own holdings, animals, and crops had not been smoke-damaged. But when Consolidated increased production, raising stacks on the zinc and lead plants in 1925 and 1927, respectively, to over four hundred feet in an effort to disperse the increased air pollution, American farmers complained. In

1924, about 4,700 tons of sulfur were being emitted each month, one ton of sulfur roughly equivalent to two tons of SO_2 gas. In 1926, this figure doubled to about 9,000 tons of sulfur a month, rising in 1930 to 300–350 tons per day—all of which for the first time caused damage over the line. Newly installed D & L treaters (dust precipitators) and the new tall stack on the lead plant resulted in much better lead recovery and abatement of a worrisome health menace to the men. But the smoke, rather than dispersing high above the river valley as planned for and anticipated, followed the river south and came to ground much farther away from the smelter. Soon "fumigations" were detected as far south as Kettle Falls, some fifty miles from the plant and well within U.S. territory.

The farmers responded by forming a committee, with John Leaden as chairman along with John Murawski and J. H. Stroh, to protest the new smoke conditions and the "invasion of our rights and homes by this rich foreign corporation" against which they had no recourse in Washington State courts.[12] All three had farms close to the line. Murawski, who had been gassed in the First World War, suffered from the fumes and soon accepted the company's offer to pay for his smoke-burned alfalfa, leaving the farm in his son's care. Stroh, with his damaged apple and pear orchard, started the farmers' protest but soon he, too, broke ranks, renting his farm to the Canadians who turned it into a model demonstration project to show what could be done in a smoke zone with modern fertilizers and good farm management. Leaden battled on as the longtime leader of the group until he died after several strokes in 1938. Negotiations with the company continued during 1926 and 1927, and some of the other farmers settled, but Leaden held the group together, forming the Citizens' Protective Association (CPA), which presented a united front of farmers and property owners. Why was he effective?

These men were of course well aware of the Canadian smoke action, recently concluded. Some of the same freelancing lawyers, engineers, and health experts hired by the Canadian farmers in their suit against Consolidated now offered their services to the Americans as they too pressed claims for compensatory damages to crops, animals, and property. The Northport area had its own history of smoke damages suits as well. But in addition to the pressure of opportunity—the motive highlighted by Consolidated as it attempted to deal with this threat from a new group of "smoke farmers"—there were economic, institutional, and cultural reasons for Leaden's persistence.

In the first of many U.S. government reports on smoke damage to this area, D. F. Fisher, a plant pathologist from the USDA's Bureau of Plant Industry, assessed the settlers' situation in 1927. During this first visit he "could discover no evidence that thus far the smoke conditions had rendered any cultivated farm on the American side unfit to grow crop," a view he would soon change. It was clear to Fisher that the main reaction was fear, "fear that they will be driven from their homes without redress, by a foreign corporation, and it is apparently this fear, rather than the amount of actual damage already suffered that is agitating them."[13]

The farmers felt economically vulnerable as well. Population was down by 25 percent, and many farms had already been abandoned before the smoke even arrived in 1926. It happened that

> the Trail Smelter damage and the hazard it creates are the chief causes within the stricken zone of the depreciation of land values, the loss of borrowing power on farms, and the inability to sell or rent them. Arrested development is everywhere apparent. Even the most progressive farmers refuse to make further outlays for improvements or to expand cultivated acreage by making additional clearings.[14]

Land held for speculation could not be sold at any price, and properties in the smoke area were not eligible for federal loans through the National Farm Loan Association.

Resistance stiffened as the company branded Leaden and the CPA as a group of agrarian radicals and agitators. Lieutenant Governor Lon Johnson, who had been the old Northport smelter's attorney, coordinated the local defense for C.M.&S. Johnson asked local business groups to pressure their town councils against the farmers and placed unfriendly editorials in the local press. Certain local opinion leaders were helped with their expenses. Consolidated also formed a competent team of U.S. academics and industry experts to work under Dr. Roy Neidig from the University of Idaho to survey the area for smoke damage. Although they found abundant past damage from the Northport smelter, there was, they claimed, little damage from the current fumigations. Instead, the visible stress on crops and forests was largely due to natural causes such as drought, beetle damage, and fire.

Angered by the editorials and suspicious of the company survey team, the settlers scoffed at these findings by men they considered hired guns. At Leaden's urging, the U.S. government sent two men from the USDA to investigate the smoke area and submit recommendations. The federal government also secured affidavits from several property owners, including a statement from the Stevens County commissioners who reported widespread smoke damage and called for action by Olympia and Washington. In small-town Northport, the smelter seemed omnipresent, from the taste and smell of sulfur in the air to the poor-mouthing and suborning.[15]

Institutional barriers now hindered a conventional settlement in what had been up to this point a localized smelter dispute. In November

1926, Blaylock, Johnson, Crowe, and the rest of their team were invited by the farmers to present the company's position. Both sides wanted to settle. However, the company was prohibited from purchasing land or holding smoke easements in the state of Washington because of the state's alien land law. When the farmers petitioned their state legislature to change the law so the smelter could pay them, they were advised by the attorney general to take up this "international matter" with the federal government. Rebuffed in Olympia, the settlers found that their leverage over the company actually increased, thanks to this institutional constraint. Some farmers settled for cash payments, but the CPA held together, and by late summer 1927 there were 115 members by Leaden's estimation.

To be sure, their cause was always conditioned by the chance that mining and smelting would take off again, in which case there would eventually be "plenty of smelter fumes evolved right here on this side of the line. I mention this," Griffin reported, "as I see the psychological effect of such a possibility already reflected in the attitude of some of the local people as to the question of smelter fumes damage."[16] In this case industry would definitely take pride of place, but the farmers living some distance outside Northport would still want a settlement.

The company offered cash settlements but would negotiate only with individuals. For starters, the evaluation placed on crop damage by company experts was a mere $1,350.75, and timber losses were said to be some $18,000. The farmers were not impressed. Crowe, the company solicitor who had dealt with the Canadian farmers for many years, and whose smelter experts under Neidig had a budget of $60,000 a year for their investigations south of the line, feigned pique at this reaction. "They will not take [our settlement offer], thinking of course that their Government will be able to get them something more than they are entitled to," Crowe wrote. He was referring to the fact that the settlers, few and seemingly so vulnerable, now had the ear of their congressional delegation.

Pressured by the state's two senators, Republican Wesley L. Jones and Democrat Clarence Dill, the State Department asked Canada to agree to refer the case to the International Joint Commission (IJC) for investigation and settlement.[17] In agreeing to take this escalating dispute to the binational level of the IJC, Ottawa asked British Columbia to step aside, even though under Canadian law the province had jurisdiction over natural resources. The Trail case was now officially internationalized, turning what had been a local dispute between a large industrial facility and nearby farmers into a more complex and far-reaching affair. That a David and Goliath dynamic was at play is clear, but when all is said and done, differences in political culture between the Canadians and Americans were a factor and worth probing as another motive for the settlers' actions.

Jack Leaden was born in 1866, the son of an Irish immigrant who had lived next door to Abraham Lincoln in Springfield, Illinois. Leaden's father formed part of a neighborhood guard that went to Washington with the Lincolns, and he later saw heavy combat as a Union army captain in the Civil War. The son went west, serving in the Texas Rangers for awhile, prospecting, then homesteading in 1901 near the border at Velvet, the farm that the family still owns and farms. He was also founder and first master of the Northport Grange, a farm improvement organization that had known its share of battles with the railroads over freight rates and commodity pricing after the Civil War but by the 1920s had long since become a conservative, grassroots, family-centered organization.[18] Politically, Leaden was a Theodore Roosevelt progressive turned socialist; his business partner later formed a successful cooperative in Spokane. He raised three sons and was a good family man. Given his politics and his Irish background, Leaden's stubborn and fiercely independent reaction to the smelter, its intrusive smoke, and its intrusive team of experts nosing around Northport is hardly surprising.

Furthermore, as leader of the CPA and the Grange, both grassroots organizations, Leaden deeply resented the efforts of Johnson and others to discredit him as a radical agitator. But as he and the other directors said in a long letter to Blaylock, "Patience was no longer a virtue. Your people," he charged, "would not recognize or deal with our organization, and considerable effort was made to block an effective organization of our people." The upshot was that unless a fair settlement were reached, "we would indeed be very sorry, for in such an event this situation might easily grow to bounds and proportions over which either of us would have but little control." That is, violence might break out. Yet he was careful to close on a conciliatory note: "We both have too much in common to wish for anything unfortunate to come to pass."[19] But the threat was there.

American political culture, so attractively responsive to grassroots politics, also contains a dark element of private violence. As a tactic, direct action is inhibiting and attention-getting. As a way to let off individual and group frustrations against government and large organizations it also has a long history in the United States. During the long investigation into Trail smoke damage, rumors of shots being fired in the woods surfaced from both sides. Starting in 1929, Leaden and his fellow settlers prevented company researchers and Canadian Government scientists from setting foot in northern Stevens County. Neidig was threatened. The company's timber expert reported that Leaden had threatened to shoot him.

Morris Katz, a Canadian government scientist from the National Research Council (NRC), was given a rough time when he attempted to argue with locals in the pool hall and was then threatened outside the

Northport Hotel. Katz and his NRC colleagues were kept out of Northport until 1934. No, the state could not send in the militia to invade a friendly country, the governor told one man at a Northport farmers' meeting in 1929. But he did understand their frustration. The settlers' mood was ugly, and they were under pressure. Blaylock received a manifesto from "a large group of sufferers" who gave him six months to clean up the smelter or die. "Do you know that you are worse than Dillinger himself?" the letter began. "We want you to spare our lives and quit poisoning the air with smelter fumes." This may have been just rough talk coming out of the culture of western mining camps, but violence was in the air and the Canadians took it seriously.[20]

James J. Warren, president of Consolidated, forwarded a photostat of this letter to the Department of External Affairs in Ottawa and a copy to the prime minister with a comment on "the obsessed people south of the line. It is written quite in keeping with the disorderly and unlawful habits of the United States citizens generally," he added. As Warren had written the prime minister in 1927, "The real trouble is not that we are doing damage—so far it is infinitesimal but that the land afflicted is not really fit for cultivation—being mostly a sandy waste along the banks of the Columbia River." Canadians were the vulnerable ones: "The unfortunate owners are hoping to extract from us what they could not get out of the soil if there weren't a smelter nearer than Mars." Thus, the company would need government help "in resisting the usual and unfair and grasping attitude of the Americans."[21]

In this way, the son of Lincoln's next-door neighbor and the president of a major CPR subsidiary squared off, the one fiercely local and independent, with support from Congress and the State Department, the other self-assured, rooted in the Anglo business culture of Montreal, and proud of a political system that concentrated power and gave him access to government at the highest levels. Both men were conservative, but in very different ways. One should not, therefore, overplay the theme of violence, real as it was.

The drawn-out Trail dispute was a lawyers' affair, in which for the amassing of evidence to assess damages scientists were relegated to a supporting role. The farmers' attorney was John T. Raftis (1892–1978), a rising star in Colville, the Stevens Country seat. Raftis had Canadian family ties; Leaden's daughter-in-law was Canadian. Like Leaden he was of Irish-American ancestry, but neither man was motivated by some sort of residual finianism. Until the mid–nineteenth century, Colville was part of the Hudson's Bay Company network of fur trading posts; Canada was part of their landscape. A graduate of George Washington Law School and having served briefly in the state legislature, he was a lifelong Republican active in community organizations such as the American

Legion, the Elks, and the Chamber of Commerce.[22] He sang in the Catholic Church choir and was an exemplary family man.

For Raftis, what became the outstanding case of his long career began as a straightforward professional opportunity: he represented the CPA and also the Upper Columbia Land and Fruit Company and a few other individuals with large holdings. As the dispute dragged on, Raftis became skilled at focusing and holding Washington, D.C.'s attention on the farmers' (and other landholders') plight in the intensely local worlds of Colville and Northport, as well as on the larger implications of the Trail case. This was the case of his life, although Raftis never reaped big fees from it.

His opponent was R. C. "Judge " Crowe, who headed the legal affairs office in Trail. A veteran litigator in smoke disputes, Crowe was a vice president of C.M.&S. The company side also relied heavily on the skills of W. N. Tilly, K.C., the CPR attorney based in Toronto and a Consolidated director. Attorneys representing the two governments were Jacob G. Metztger of the State Department and John E. Read, legal adviser to External Affairs and later a member of the International Court of Justice.

Here then were the players in what had become a transnational drama of extraordinary interest. Canada had the giant smelter but was inexperienced in foreign affairs, this being its first major dispute to be handled outside the British Foreign Office. The United States, hugely bigger by comparison, had the task of defending a handful of stump farmers, or smoke farmers, depending on where one sat. O. D. Skelton, the undersecretary of state for External Affairs, summarized the opposing forces for the prime minister:

> [The farmers see themselves as] honest and struggling, home-loving, peaceful farmers in Washington being driven from their happy homes by a greedy and arrogant alien corporation. The Company views the case as an attempt at holdup by farmers in a nearly hopeless section who have come to think they can get much more out of farming this rich corporation across the boundary than from farming their farms, and who are endeavouring to use the Governments at Washington and Ottawa to threaten a complete cessation of operations and thus force extravagant indemnity. There may be some truth in both views, but from what I have been able to see of the situation I think the Company's case is much the stronger.

For their part, U.S. officials, including USDA scientists sent to the field in 1927 to investigate the smoke complaints, sympathized with the farmers. "I have a great deal of feeling for these most unfortunate people," wrote M. C. Goldsworthy, a young plant pathologist, to his bureau chief.[23]

These were, after all, desperate times. Some 597 people, or 129 families, lived in the district of greatest fume damage, the county relief supervisor reported in 1934. "These people realize that their condition is not the result of being on sub-marginal land, but is directly the result of the greed of a foreign corporation, which if ended by the cessation of the fumes would restore their property." In dire straits, they were determined to stay the course while "their organization [the CPA] is now pressing for a solution." Their CPA, their political leverage, and consequently their ability to focus Washington's attention on the possibility of a big settlement gave them influence out of all proportion to their numbers or economic status.

During the Depression's worst years, Trail suffered from production cutbacks, although Blaylock managed to keep the workforce employed. The stock market crash of 1929 came just as Trail was beginning construction on a $9 million acid recovery and fertilizer complex, which the company claimed would solve the fumes problem.[24] Much of the investigation by both sides took place during the depths of the Depression, which undoubtedly sharpened grassroots perceptions of the conflict.

At this point in the narrative it may be wondered why an industrial problem so local in its causes and effects was not handled jointly and cooperatively by the authorities closest to the problem, as it would be today. The short answer is that because of the transboundary aspect, Washington State officials were only too eager to turn the matter over to the federal government. "In the fall of 1927," Henry Knight of the USDA recalled, "the people in the area around Northport, . . . the officials of the [State] Agricultural Experiment Station of Washington, and the Governor pleaded with Congress and the Department of Agriculture to have some action taken." In decentralized Canada, where the provinces had jurisdiction over agriculture and natural resources, External Affairs had wanted British Columbia to handle it. The province did conduct an investigation but withdrew when the Americans asked that the IJC take responsibility for an investigation and settlement. This action federalized the issue.[25] Many years later, in 1991, the two subfederal governments set up an Environmental Cooperation Council to deal with a range of transboundary issues, including pollution from the Trail smelter.[26] But in the 1930s, this harbinger of "the new regionalism," or cooperation across borders at the subfederal level, was not an option.

With the state's two U.S. senators and Sam Hill, the local congressman, clamoring for relief for the settlers, opening a reference with the IJC was acceptable, both to James Warren of C.M.&S. and to O. D. Skelton, the senior civil servant who as undersecretary of state at External Affairs coordinated the Canadian position. "The Commission is a competent and fair body and well adapted to making such an enquiry," Skelton wrote in late 1927. "As a matter of general policy, it is the Prime Minister's view that it is to Canada's advantage that every legitimate use should be made of this agency of conciliation, from which Canada as the smaller country stands in the long run to gain more than the United States." All the same, the Canadians were reluctant to agree to terms of reference that would in any way "appear to endanger the existence of the Consolidated plant at Trail," especially as "the majority of the stock of this important company is owned by the CPR, which is the most powerful commercial activity in the Dominion."[27] From the Canadian point of view, federalizing the dispute under the IJC helped to equal the sides.

Pressed by the Washington State congressional delegation, the State Department prepared and coordinated the U.S. case in support of the settlers. But State and the Department of Agriculture both had more encompassing reasons for getting into the dispute. In requesting that the Bureau of the Budget approve the then large sum of $40,000 to support a thorough scientific investigation by scientists from the USDA, the State Department said it "was not interested alone in the matter of damages in this instance. It was interested in the principles providing a remedy for losses suffered by property owners in this country from acts committed by corporations located across the border in contiguous countries." Thus Washington was looking for more than a remedy to this specific case at issue. Canada was probably the main target, for in later years transboundary pollution from Algoma Steel and the INCO nickel works at Sudbury, Ontario, came up in discussions between the two governments. But the fact that claims for damage to property on the other border stemming from the Mexican Revolution were also being negotiated may have been a consideration. Jacob Metzger, the State Department's legal adviser who coordinated the Trail smelter case, was a veteran negotiator of claims with Mexico.

For its part, the USDA, at the Bureau of Chemistry and Soils under Dr. Henry Knight, was ready to do scientific battle with the smelters. Having been through a generation of smelter disputes affecting agriculture and the national forests—Anaconda, Ducktown, Selby, Salt Lake, to mention the most prominent—the department was eager to protect the Stevens County farmers in this instance while using the investigation to establish a new, general standard for pollution control. It was not long before Knight and his colleagues were talking about publishing a four-hundred-page report that would replace the Selby Smelter Report as the industry standard.[28]

D. F. Fisher, the head plant pathologist, and J. S. Boyce, a forest expert, arrived in Northport during the summer of 1927 to conduct an independent reconnaissance while conferring with the company's team under Roy Neidig. In the summer of 1928, W. W. Skinner, the chief chemist, surveyed the area and developed a plan of work that he coordinated through an informal interagency group called the Smelter Fumes Committee. A veteran of many battles to enforce the Food and Drug Act, Skinner was a water standards expert who lived by the creed "thou shalt not pollute." Chaired by Knight, the Fumes Committee convened in May 1928 and met for the next ten years to plan and evaluate Skinner's research effort. K. F. Kellerman, head of the Bureau of Plant Industry and a veteran in smelter disputes, attended. So did Jacob Metzger, the State Department's legal counselor, for whom the scientific evidence was being gathered. As both paymaster and coordinator of the litigation, the State Department ultimately called the shots.

To establish its own independent assessment, External Affairs directed the National Research Council of Canada to send a scientific team to Northport and in 1929 set up the Associate Committee on Trail Smelter Smoke under the NRC. John E. Read, who was Metzger's counterpart at External Affairs, sat on this committee, which was chaired first by H. M. Tory and later by General A. G. L. McNaughton. At first this research was funded by the NRC, but as will be shown in Chapter 3, Consolidated became the paymaster and through John Read called the shots for Canada. Thus, within a few months of one another, each government had an interagency smelter committee to prepare its side of the scientific investigation for the IJC.

Just as the Americans through their field investigators maintained close relations with Jack Leaden and the farmers, so too did the Canadians work closely with Consolidated's attorney R. C. Crowe. Both teams presented evidence before the IJC commissioners under the supervision of their respective legal advisers, Metzger and Read. In turn, the IJC hired experts on its own behalf to evaluate the work of the government scientists: Dean E. A. Howes of the University of Alberta and Dean F. G. Miller of the University of Idaho. Meanwhile, the original company team under Roy Neidig continued to operate the experimental farm on Stroh's leased property and were active south of the line until the settlers ran them off in 1929.

The farmers' ordeal by experts had only just begun. Stewart Griffin, the USDA's head chemist based in Northport whose sympathies were with the settlers, captured the scene a full decade later. Emissions, though by that time much reduced, still threatened the farmer's peace of mind, and he was still being "over run from year to year by 'experts' who at the best are seeking for scientific facts and at the worst are looking for evidence to be used either for or against him. In the meantime, the knowledge that his property is in the so-called 'smoke' area, has destroyed his chances for sale or for a reasonable cash rental."[29] Trusted by the settlers, USDA scientists developed close friendships in the Northport area but were also expected to deliver the evidence on which a large damage award could be based, a factor that affected how they saw their mission. From the start they were placed in an awkward situation: on the one hand, they were working in the interests of the farmers, "but on the other hand," Knight wrote in his diary, "we are working for the Joint Commission." Furthermore, the IJC, an impartial international body, would have the benefits of their research, but the Smelter Fumes Committee, to which they reported, was charged with gathering evidence to be used by one side of the dispute.[30]

With all this investigatory activity in northern Stevens County, it is clear from the record that official assurances of "reasonable cooperation" among scientists in the field soon gave way to competition and actions more appropriate to litigants in a dispute than to an impartial scientific investigation. Soon after his arrival in Northport, Griffin asked Skinner for instructions about cooperating with Neidig. "In regard to joining forces with Professor Neidig in making simultaneous investigations of SO_2 in the air of the Columbia Valley, I can appreciate strong arguments both for and against," his boss replied. If Griffin were to join forces with him, "which undoubtedly would be economical and most satisfactory, because I have a high regard for Neidig's ability and integrity," he would have to share results with him. This collaboration would "immediately jeopardize the sympathetic support of all the people affected in that region," who saw Neidig, Johnson, and the others as bought "by the company's bag of gold." In other words, the settlers would not understand or support him. Finally, Skinner concluded, there was "this other consideration, which you must always keep in mind; you are in fact obtaining evidence to be presented before a court. In this I have had quite a good deal of experience and I have always found it to be the best policy not to permit your opponent to prepare his rebuttal material in advance of your presentation."[31]

Griffin should be clear about his mission: it was not to assist the smelter people in solving their emissions problem, a position for which the IJC commissioners developed strong sympathy and because of which they had requested all the experts to cooperate on finding a solution. Rather "we must have chemical evidence to show the extent of the distribution of the fumes on this side of the boundary. Everything else is subordinate to that one thought."[32]

Little wonder that Warren wanted Canadian government scientists in the field. "We cannot feel safe or secure without Canadian witnesses of the same independence as the United States federal ones. Some of the latter seem fair and unbiased but others are not, notably Skinner. He is a partisan and a prosecutor rather than a witness." The United States is "assembling all kinds of testimony which if wrong and we have no witnesses to correct will work irreparable injury," Warren concluded.[33] The Canadians were also put off by the adversarial tactics employed by Metzger in the IJC hearings, tactics that instead of furthering cooperation reminded them of behavior by "a small town police lawyer." As a subscriber to the lawyer's creed that it was not his job to facilitate an opponent's case, Metzger might have phrased it somewhat differently.

Charged with gathering and evaluating evidence for the IJC, Deans Miller and Howes asked Griffin for a map showing the extent of the smelter fume zone for use in preparing their own report. Instead of giving them the map, which was still being elaborated, Griffin decided "to furnish them with the results we obtained at certain definite points in which they were interested." He added: "I hardly know what course to pursue in this matter and hope you [Skinner] will not consider that I have acted unwisely." Dr. George H. Duff, the head plant pathologist at the NRC, asked Griffin to help with building and calibrating the new SO_2 monitoring machines that the Canadians were setting up at Marble and near Northport in September 1929. Since both the Canadians and the Americans bought their recorders from the ASARCO research station in Salt Lake, Griffin complied. He also compared notes with Morris Katz, the NRC's head chemist in the field, and when Katz's data on the effects of low concentrations of SO_2 on plants seemed to confirm Griffin's findings, Skinner was delighted.[34]

But everything was geared toward the damage award, while simultaneously doing good science to support the case. Herein lay the subtle gap between two different definitions of the mission: "Rather than approaching the problem from the standpoint which the Smelter Company has attempted, whereby certain damages are classified as economic and others disregarded, it is essential to evaluate all types of damage."[35] To the Smelter Fumes Committee, this situation merited a full and comprehensive scientific investigation in addition to an economic survey of actual physical damage. But from the company's point of view, the Americans were conducting their work with an effort to find damage rather than to prove if damage was actually being done. Although this interpretation was not the whole story, it was useful for discrediting the U.S. research effort in the next set of hearings, as will be discussed later. For their part, the Canadian government scientists would be drawn toward this narrower interpretation of their mission.

In 1929, while wanting an independent investigation of their own, the Canadians said they would appreciate the opportunity for cooperating with U.S. scientists already in the field, and the USDA instructed its scientists to extend full cooperation. It is important, Skelton wrote in 1929, to get scientists from both governments "to reach some common ground."[36] Just how difficult it was to accomplish this goal was revealed at the IJC's public hearing held at Nelson, British Columbia, in November.

The United States had been expecting a settlement, but at the last minute the Canadian section pulled back at the company's request, and the hearings went on. Consequently, the State Department's Metzger did not want to reveal the U.S. position prematurely and instructed the USDA team "to withhold everything," Griffin reported.

> The Canadian experts were there in force and we all felt that it would indeed be unfortunate if we were to reveal all our findings and the Canadians were then allowed several months to prepare for rebuttal. When Mr. Metzger ascertained that this was the strategy arranged, he declined to have us go on unless rebuttal was immediate. Moreover when it became evident that the Canadian scientists would like to draw us into an informal discussion in the late afternoon and evening, Mr. Metzger suggested that we avoid this by leaving town immediately as there would be no further sittings of the Commission at this time.[37]

There was nothing subtle about Metzger's approach. Otherwise, informal discussions between scientists would have furthered their ability to pinpoint, evaluate, and advance a common research agenda.

Perhaps even more revealing was the reply George Hedgcock, the USDA's senior forest pathologist, received after asking his boss whether to cooperate with Andrew McCallum, his Canadian counterpart, who requested it in 1930. Skinner had advised him not to cooperate, but Hedgcock's bureau chief had said to use his own judgment:

> I feel there is a very important psychological advantage in agreeing to cooperate in work about which there may be controversy. The cooperative work is less subject to criticism and has the additional advantage of furnishing some warning of the kind of work that is being done by the opposing forces. I think your work will not suffer in any way by them understanding what you are doing.[38]

The record shows clearly that time and again the Canadians with three men in the field took the initiative in asking for cooperation, while the Americans with as many as eight researchers during this early phase of the investigation held back—not that they wanted to as scientists, but that they felt they ought to as seekers for evidence to support the farmers.

In 1932, months after the IJC had finished deliberating and had issued a ruling quite favorable to the U.S. side, the Canadians suggested that "our experts should exchange the scientific data which they are collecting in connection with the Trail Smelter case." But with the farmers opposed to the settlement, and Congress loath to appropriate more money for research, attitudes hardened. Until the future course of the investigation and the funding situation was clarified, State replied, "it does not seem feasible at this time to make arrangements for the exchange of data." Rather than cooperating, the two scientific teams became polarized.

Later, the international Arbitral Tribunal that oversaw the final stages of the dispute was sharply critical of the two sides' failure to pool information and carry out joint investigations. In 1938, a suggestion to do a joint crop survey was turned down flat by Griffin:

> Aside from any ethical considerations involved there are practical reasons that would prevent me from going into every area under conditions which would frustrate the accomplishment of my mission, and I feel that I should now make it clear that I cannot attempt the assignment if I am not to proceed independently of the Tribunal or of the Canadian smelter technicians.

Uppermost in Griffin's mind was retaining the trust and friendship of local citizens. "Several of the farmers sought me out [during the earlier IJC hearings] in order to express their appreciation. This was more or less of a satisfaction to me, I will admit," he reported back to Skinner. Having lived in Northport off and on for almost ten years, "Griff" and his wife, Doris, received lots of sympathy when their baby was stillborn. Leaden's son Howard assisted in Griffin's and Hedgcock's laboratory. But Griffin was also painfully aware that the two scientific teams had long since diverged in their assessment over what sort of damage had actually occurred, and he vigorously disagreed with consultants for the tribunal who claimed no damage was occurring in 1938, considering them pro-smelter. For its part, the tribunal referred several times in the second set of hearings to this "lack of cooperation . . . as being a prominent feature of this case." And "after four years [since hearings began]," the chairman said pointedly in 1940, "the two governments have not even agreed on a common tabulation of recorder data."[39]

In smelter litigations it was the usual practice for both sides to hire their own experts. It would be easy to agree with the *Globe and Mail*'s mining editor who commented that in litigation of this sort the "experts in every line usually give evidence that is valuable to the side that employs them."[40] But these were qualified government men, public servants, and scientists being pulled into adversarial roles because their governments had divergent goals.

Since 1928, the Americans had mounted a major effort to prove that forests, vegetation, and crops were subject to a chronic and cumulative sort of injury from low-level SO_2 fumigations that could not always be detected by burn markings. This contention was known as the "invisible injury" thesis, and optimism ran high on the USDA team that they could prove it. As D. F. Fisher, the senior plant pathologist, reported to his bureau chief: "Our preliminary work indicates that there is a chronic type of injury to plants from repeated fumigations with low concentrations of SO_2. This, of course, is not accepted by the smelting interests and there is very little corroborative evidence in the literature."[41] To abate such injury—if it really was chronic and cumulative at low intensities, as the Americans maintained—the Trail smelter (and indeed the whole industry) would be required to establish a very restrictive operating regime or even, in the case of certain smelters, to shut down.

With their industry on the line, NRC scientists were certainly aware that results upholding the traditional thesis, namely, that "visible injury," or the effects of outward markings at certain intensities on foliage, was the only credible standard for assessing damage, would be welcome. Consolidated and the government wanted a credible, independent team to monitor the work of USDA scientists and to draw conclusions from their own field research on the nature and extent of crop and forest damage. The NRC results were favorable to the industry position, and the nicely managed Canadian research effort produced a set of important papers that was published in 1939 and became a mainstay of the air pollution literature for several years.[42]

For both sides, the pressure to give evidence that would stand up in court did channel results and blur uncertainties. However, being in a long and complex litigation also gave the scientists the time and motivation to conduct innovative research in a highly complex field. Sixty years later, experts in what is now called the science of acid deposition largely agree on the phytotoxic effects of the damage being caused by SO_2. Yet for all the money, time, and effort expended, the scientific results of the Trail investigation were controversial as well as underwhelming, as will be discussed in the next chapter.

By contrast, little controversy attached to the masterful survey of forest damage by George G. Hedgcock (1863–1946), the senior Forest Service pathologist. A veteran researcher who had presented evidence of forest damage in two earlier smelter cases, Anaconda (1910–1911) and Ducktown (1914), Hedgcock's methodology and findings were accepted by the IJC and carried great weight in the assessment of forest injury and the award for damages handed down in 1931. The Citizens' Protective Association trusted him, and when Hedgcock was seen walking through the forest in the high boots and khaki uniform of an official in the U.S. Forest Service, this elderly, rather pudgy man with his grandfatherly demeanor was a reassuring presence to the angry settlers.

Having first distinguished smoke damage from winter and drought injury to ponderosa pine, Douglas fir trees, and forest shrubs, Hedgcock quantitatively delimited and mapped the geographical distribution of SO_2 injury in three zones. This map showed that the southern limits to injury were some fifty miles south of the line, nearly to Kettle Falls. The zone of greatest injury lay near the border, where the valley was comparatively narrow and deep, at times acting as a flume, confining and guiding the passage of sulfur dioxide released in it. In addition to burn markings, injury was identified by evidence of retarded growth and the reduced rate of seed cone reproduction. Based on sampling techniques, Hedgcock could state with a fair degree of certainty that forest growth and seed cone reproduction had slowed after the high stacks went in, and where this injury had occurred.

Hedgcock's major fieldwork extended from 1928 to 1931, his map and report being prepared when peak outputs from the smelter averaged about 9,300 tons of sulfur a month (during 1926–1930). By 1931, the combined effects of production cutbacks because of the Depression and the installation of sulfur recovery measures cut emissions to about 7,100 tons a month. Additionally, in 1933 the smelter imposed an emissions regime to regulate sulfur discharge so that the larger amounts would coincide with wind direction and other atmospheric conditions least favorable for SO_2 injury.[43] As emissions fell irregularly thereafter, Hedgcock observed that the rate of reproduction among ponderosa pines improved; although injury still occurred as late as 1936, a new map showed significant reduction in the affected zones, which were reduced to two.[44] By 1949, his associate (Theodore C. Scheffer) observed that "trees even in the northernmost part of zone 1 were in excellent condition and apparently had not been importantly affected by sulfur dioxide since 1940 at the latest."[45]

With its precision and lack of ambiguity, Hedgcock's map was very well received by the IJC commissioners. The findings of both Griffin and Katz on the limits of SO_2 damage corresponded well with it. The IJC damage award of $350,000 was largely based on observed economic injury occurring in 1930 in Hedgcock's three zones, comprising some 102,290 acres. The total U.S. damage claim of some $700,000 included indemnification for depreciation on properties whose sale value was lowered by virtue of being in a smelter zone. The Canadian commissioners would not go along with this argument, a strong blow to a main element in the farmers' claim. That the depressed land values prevailing *before* the stacks were raised were also factored into the award was another blow to the farmers. No damage was assessed for losses in tax revenues to Stevens County and the City of Northport, nor was damage assessed for alleged ill health to individuals.[46] Although the farmers were bitterly disappointed, the award was almost six times greater than the Canadian farmers had received some years before. Consolidated, whose offer of $250,000 had been rejected by the farmers, regarded the IJC award as too liberal.

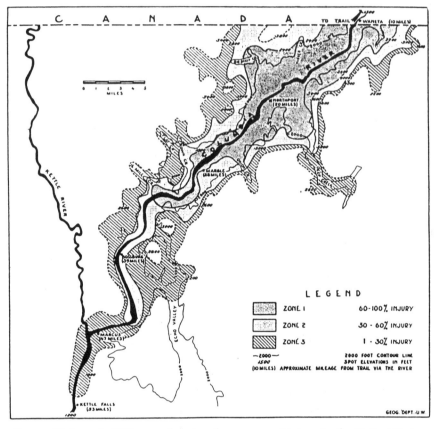

George Hedgcock, 1929 map showing three zones of injury in the United States from Trail smoke. (U.S. National Archives)

Equally well received by the IJC was Blaylock's promise to abate emissions by 30 to 35 percent through an aggressive acid recovery program to the point where SO_2 emissions would do no appreciable damage in the United States. Since 1926 the company had embarked on a research program to prevent or considerably reduce the smoke problem. A new process was found for removing substantially more sulfides from concentrates at the Sullivan mine, and a new method for burning concentrates in the zinc roasters was developed that enhanced SO_2 recovery. (A contact acid plant requires gas of at least 4 percent and preferably 7 percent SO_2, which the new process delivered.)

The more difficult problem of cleaning up the gas from the lead smelter was addressed by developing an acid absorption process. A small acid plant was operating by 1929, the beginning of a $9 million construc-

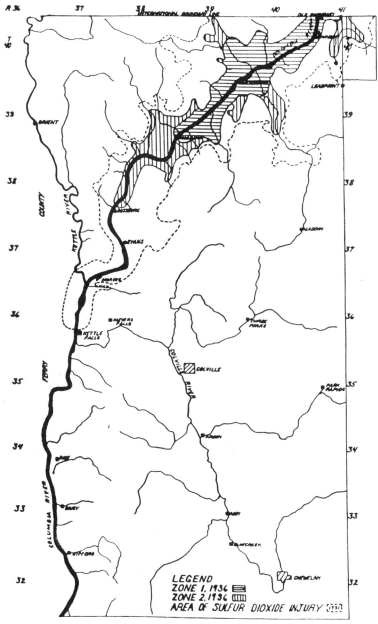

George Hedgcock, 1936 map showing a substantially improved landscape. (U.S. National Archives)

The Trail smelter, 1929. (U.S. National Archives)

tion program to build three larger acid plants. The sulfuric acid was then used in the nitrogen fixation program to produce ammonium sulphide for fertilizers. Down the road plants to recover elemental sulfur and produce superphosphate fertilizer would be built.[47] Under the direction of Roy Neidig, the company's Elephant brand fertilizer was marketed to the network of prairie farmers developed by the Canadian Pacific Railway in California, and as far away as British India. Using what had been a difficult to market and unwanted by-product, sulphuric acid, this fertilizer became a profitable part of the business for several years until the 1970s.

In order to gradually reduce pollution, Blaylock also planned to phase in an economically viable control process. In 1933, a smoke control regime was installed to limit the duration of burns (the farmers' demand for the immediate cessation of all smoke damage was not feasible from a business point of view). The problem centered on the occurrence of occasional acute burns. The atmospheric conditions were such that a stable inversion layer frequently formed at night, trapping all the stack gas in the valley along which it spread either north or south or both, depending on the wind. At night, little SO_2 reached the ground, but in the morning, as the rising sun warmed the west side of the valley, the air formed a slow vortex about an axis along the valley's middle that "caused a high concentration of sulfur dioxide to reach the floor of the valley with severe damage to all of the vegetation."[48] In addition to improvements at the

George Hedgcock, U.S. Forest Service, Senator Clarence Dill, and John Raftis (1937?). (Courtesy of Walter Raftis)

plant, part of the solution was an innovative SCS regime, developed in 1938, as discussed below. By 1940, the percentage of SO_2 being removed from flue gas had risen to 70 percent, the rest being vented into the atmosphere. With the plant running at full capacity for war production, this was a major achievement.

Among North American smelters, Trail took the lead in efforts to abate pollution, not simply in reaction to litigation, but as the result of policies to promote better industrial management practices—developing technical innovations to reduce sulfides at the mine and SO_2 at the stack, and finding new markets for the by-products. Faced with the need to

Selwyn Blaylock, superintendent of the Trail smelter. (Courtesy of COMINCO)

abate pollution, Consolidated turned a threat into a technology-forcing opportunity, becoming the industry leader in the process.

"The age of by-products [in smelter production] has begun," proclaimed Robert Swain, a professor of chemistry at Stanford and perhaps the best-known expert on smelter pollution in the United States. "The old age of litigation between industry and agriculture could now be replaced by cooperation, to everyone's benefit," he concluded. G. L. Oldright, the supervising nonferrous metallurgist at the U.S. Bureau of Mines and a frequent visitor to Blaylock's domain, expressed his admiration in public at the IJC hearings and privately to Trail management. Writing to R. W. Diamond, the company's chief chemist, Oldright praised "the courageous pioneering work being undertaken by the Consolidated Company to solve the whole question [of smoke damage] by the use of engineering methods in such a way that the net result would be to benefit the world at large." As for industry, this development "cannot fail to meet with the interest and the sympathy of all engaged in similar pursuits," he added.[49] In fact, Trail management never lacked for American experts to consult, advise, and cheer on their efforts to find technical solutions to this very difficult problem.

The upshot is that the IJC sanctioned an innovative technical and economic solution to the fumes problem, but it was one that would reduce, but not eliminate, injury to American property holders—a U.S. demand that was watered down in the final terms of reference to the IJC, which contained no express provision for eliminating injury. As Skelton wrote to the prime minister in 1928, "We can rest assured that there will be no unanimous recommendation from the Commission to that effect unless there is overwhelming evidence for it."[50]

While the scientific evidence for plant damage was less important than Hedgcock's findings in determining the IJC award, by 1930 both of

ELEPHANT BRAND
FERTILIZERS

Ammonium Phosphate — Triple Superphosphate
Ammonium Sulphate

ELEPHANT BRAND
Fertilizer Feeder Attachment

An Attachment Designed for Successful Fertilization
to Suit ALL Makes of Seed Drills

Manufactured at Tadanac, Trail, B. C., by

The Consolidated Mining and Smelting Co.
of Canada, Limited

CHEMICAL AND FERTILIZER DEPARTMENT

Head Office: TORONTO GENERAL TRUSTS BLDG
CALGARY, Alberta.

Branches:

Hotel Saskatchewan   212 Confederation Life Bldg.
REGINA, Sask.   WINNIPEG, Man.

Elephant brand fertilizer pamphlet cover. (Courtesy of COMINCO)

the parties believed they had reached important results even if the commissioners themselves were puzzled by the contradictory evidence being
presented to them. "What we had established," Henry Knight, chairman
of the U.S. Fumes Committee, reported in his diary, "was that there was
damage and serious damage in the area around Northport due to smelter
fumes; further we had established the fact that this damage was accumulative." Dr. M. C. Goldsworthy, a pathologist with the Bureau of Plant Industry, wrote of "the real progress our research program is making
towards finding out the solution of this most perplexing problem." If the
award itself seemed small when compared with the original claims, "we
were quite happy to hear of the award and think it a great victory for our
side." For their part, the Canadians had reached quite different conclusions, based on the results of *their* research. As Skelton informed the
prime minister, "Evidence collected independently by the Dominion
Government through the National Research Council did as a matter of
fact largely confirm the evidence presented by the Smelter Company's
witnesses, who were chiefly United States scientists."[51]

Secretary of State Henry Stimson was inclined to go along with the
award, even though it was only half of the $700,000 the United States had
asked for, and Jacob Metzger, the legal adviser, got him to modify his favorable public comments on the award. Nonetheless, Griffin reported
back to Skinner that

> in respect to the amount awarded, Mr. Metzger feels that although it
> falls far short of the claims submitted, nevertheless it is a very sub-
> stantial sum—particularly when viewed in comparison with the few
> thousand dollars which the smelter would have perhaps made
> available to the farmers should no pressure have been bought to
> bear. Mr. Metzger and Mr. Murdock [the Department's other lawyer]
> view the agreement with much gratification as a moral victory for us
> and a vindication of our viewpoint in respect to the scientific part of
> the investigation.

Because Metzger left few records, Griffin's testimony is important for establishing the actual role that the scientific evidence played in Metzger's
thinking. He continued:

> That part of it, I believe, is the feature which tempts them to consider
> acceptance of the plan. They believe that a settlement, based on even
> $350,000 would be a fitting capstone for our scientific efforts and
> achievements, such as they are. He considers that our report when fi-
> nally compiled (and supported and strengthened by a substantial
> award to agrarian interests) would be of value as a precedent in the

consideration of other similar problems, as they might arise in the future.[52]

Recall that, in backing the Stevens County farmers, both the State Department and the USDA had larger goals: once published, the scientific report would be useful in future transborder claims and also for setting a new industry standard for smoke abatement.

Not that the settlers' cause took a back seat to these concerns. Goldsworthy, the personable plant pathologist known as "Goldy"—who "had a great deal of sympathy for these most unfortunate people"—thought "the award should go a long way towards alleviating the conditions of the settlers in the immediate vicinity of the border where the damage is quite large."[53] But acceptance of the award was put on hold, most likely because of concerns about the progress of negotiations over the St. Lawrence Seaway Treaty. Nothing happened until the farmers forced Washington's hand in the last days of the Hoover administration.

Bitterly opposed to the settlement, Leaden and his colleagues wondered why, if evidence for invisible injury existed, the award was so low. They also objected to the Deans' Report, prepared for the IJC, which recommended that the whole Northport area be declared an industrial zone on the grounds that it was obviously much less suited for agriculture than it was for mining and smelting operations. In fact, the deans played into the hands of industry spokesmen who portrayed the area as an underdeveloped wasteland.

With Senator Dill actively supporting the farmers, and with Congressman Hill and the governor of Washington also demanding redress for the settlers in strongly worded letters, the State Department in 1933 officially reopened the case. The formidable William E. Borah of Idaho was "on the warpath." As a longtime member of the Senate Foreign Relations Committee, his grandstanding on the settlers' behalf did not escape President Roosevelt's attention: Borah had the power to disrupt the New Deal, and he was adept at promoting western interests. The president became actively involved in 1934 and urged action toward a settlement.

As pressure mounted, Metzger and John Read met with representatives of the claimants and the company in Spokane. Instead of accepting the IJC award, Washington was now talking of reopening the whole question of damages from the beginning, and "their proposals for future restrictive measures are now quite stringent," Skelton wrote in October. The company was facing a "very difficult and dangerous situation," Warren wrote to Read. Having spent $100,000 on the first hearings, the prospect now was to face a second set of hearings that could well result in additional costs and a much higher award.[54]

Furthermore, the question of a much more stringent regime to reduce smoke damage than the company intended was now raised by the State Department, which was what Metzger and the USDA had wanted since 1927. However, no doubt for tactical reasons, the company portrayed this restriction as something radically new. "At no time in the past history of the Trail case . . . has the issue of determining the maximum frequency, duration and concentration of Sulphur Dioxide visitations which can be permitted in the State of Washington, been the case," Crowe explained to John Read. He continued:

> Before, it was a case of assessing what damage had been done. Now, a new issue is raised by the United States . . . [which] brings in a new principle not only in the Trail Case but in practically all smelter smoke litigation, if not all, namely that of setting a rule as to the frequency, duration and amount of concentration permissible without doing damage or injury. The factors that would have to be considered [in addressing this issue] (being concentration etc. of sulphur dioxide, humidity, light, temperature, the kind and stage of growth of vegetation) can be reproduced in any locality where sulphur dioxide is present.[55]

Consolidated's considerable influence in Ottawa was apparent when language referring to an operating regime for the smelter was softened in the reference, leaving it to a three-man arbitration panel to decide whether, if damage had been caused since 1931, Trail should refrain from causing further damage and if so, to what extent.

Such was the power of the Citizens' Protective Association that Metzger met with them in March 1935 to discuss the terms of the upcoming arbitration, including the Canadian language that could well lead to the tribunal's finding that some continued pollution would be necessary instead of the outright prohibition sought by the settlers and the government. John Raftis, the farmers' attorney, had already convinced them that it was in their best interests to be less intransigent. As a result, the CPA allowed three Canadian scientists to enter Northport in June 1934, where they resumed fieldwork with the governor's blessing. At Metzger's urging the CPA resolved "to accept the proposed convention rather than allow the matter longer to remain in a state of uncertainty."[56] These terms were then agreed to by the Canadian and American governments and ratified by the U.S. Senate in June 1935.

Held under the auspices of the Trail Arbitral Tribunal, composed of a three-judge panel, the next set of hearings was a great disappointment to the Americans and a triumph for C.M.&S. On June 22, 1937, the tribunal members held a preliminary hearing in Washington, D.C.; the three judges visited Northport and Trail from July 1–6 and then moved on to

Spokane for the first substantive hearings, which lasted until July 29. In Spokane, the American scientific case fell apart, and the farmers, facing defeat, bargained furiously with Consolidated behind the scenes but came away knowing they would receive next to nothing from the new set of claims. Before recessing, the tribunal met next in Washington, D.C., from August 16–19 to read field reports. A ruling on the extent of damages and an award were then issued in April 1938.

As expected, Consolidated was assessed a very small penalty for polluting, only $78,000 for damages sustained from 1932 through 1937. American attempts to widen the scope of damage to include depressed real estate values and the consequent decline of the local tax base at the town of Northport and vicinity were turned aside. Expenses for carrying out additional scientific work since 1931 were disallowed, along with court costs and a roundhouse claim for injury to U.S. sovereignty. Given that the IJC award had not yet been distributed, the long-suffering claimants and their government received very little for their efforts in this second round of hearings.

Moreover, the assessment of injury was limited to include only visible damage to forests and crops.[57] Attempts by USDA scientists to show that there had been long-term, cumulative (or "invisible") injury by SO_2 to forests, crops, and the soil were effectively excluded or discredited in the public hearings at Spokane. During the IJC hearings in 1929–1930, the Canadians had been taken aback by the aggressive tactics of the Americans, but in Spokane that summer of 1937 the shoe was on the other foot. Lawyers for Consolidated aggressively demolished the U.S. scientific testimony while ridiculing the methodology developed by R. T. Sanders, the USDA's agricultural economist, to estimate property damage. The voluminous transcript leaves no room for doubt that Canadian scientists arguing against the invisible injury thesis performed better than their American counterparts, and that Sanders himself did poorly.

"Judge" Crowe and John E. Read stayed with the case for the duration and developed it in a close and effective partnership between industry and government.[58] Pitted against Crowe and Read was Jacob G. Metzger, an able State Department attorney with considerable experience negotiating international claims, especially with Mexico. Somewhat of an idealist and deeply committed to the Stevens County farmers, Metzger was pressured by U.S. smelter interests who, when he would not bend, tried to have him fired. Unfortunately, Metzger had been carrying the U.S. argument in his head, so that when he died suddenly in early 1937 leaving no written game plan, the American scientists went into the hearings that summer in Spokane unprepared and ripe for defeat.[59] Although John Raftis took over the case in 1938 as the U.S. agent and did well in subsequent hearings, he was outclassed by the Canadian legal talent; and while Raftis had the weight of the U.S. government behind him, that government was divided, even disengaged.

Moreover, the Canadian side had the backing and support of American smelting interests, who not only shared information on their own smoke abatement technologies but also and more importantly contributed the results of their scientific studies. By contrast, the United States could not even muster a united front. That the USDA scientists had scores to settle with the smelters is clear. But officials from the U.S. Bureau of Mines (USBM), brought in for technical advice on fume remediation, had just the opposite motivation. The USBM saw its mission as service to industry. For years the bureau had been working with the smelters on ways to enhance recovery of by-products and to effect better smoke dispersion at the stack. The USBM approach was to address the technical aspects of the problem rather than investigate the chemical effects of air pollution. It may be surmised that their training as engineers predisposed them to a narrower interpretation of the public good than that of the biologists, botanists, and chemists of the USDA.

At the IJC hearings in 1929, the USBM's Dr. G. L. Oldright had reviewed Trail's abatement program, including the acid plants and fertilizer production plans, and declared it to be outstanding and state-of-the-art. Whatever SO_2 remained after treatment would have to be vented, but damage (if any) to farmers would be amicably resolved, which was standard industry practice, he told the commissioners. In praising Trail's pioneer efforts at control, Oldright helped to legitimize the Blaylock program, and the hard-driving superintendent often pointed this out. Oldright's testimony "didn't do the farmers any good," observed the USDA.[60]

Dr. Reginald S. Dean, chief engineer at the USBM Metallurgical Division, was a highly respected technologist and expert on smelter smoke dispersion who took a dim view of the USDA effort in general and of the State Department strategy of confrontation as orchestrated by Metzger in particular. Instead, Dean advocated a cooperative approach with industry to obtain a technical solution to the SO_2 "nuisance" and was delighted when it was announced that a metallurgist rather than the foods chemist nominated by the USDA would be appointed U.S. scientific adviser to the Arbitral Tribunal. This appointment had the blessing of Swagar Sherley, Metzger's lackluster replacement as U.S. agent. Evidently coached by Dean, Sherley

> was reported to be opposed to this [USDA] selection and feels the American arbitrator [on the Tribunal] should have the advice of a metallurgist so that he will not hold out for some operating requirement being placed on the Trail smelter which might at a later date

seriously embarrass the domestic smelters, particularly in a controversy with Mexico where the shoe would be on the other foot.

Of course, this was the industry position. Soon Dean himself was appointed to the post of U.S. scientific adviser to the tribunal.[61] Chemistry
professor Robert E. Swain of Stanford was appointed scientific adviser to
the Canadians, having been convinced to take the position by the president of ASARCO.

Bureaucratic infighting is a feature of U.S. divided government, and
the USDA came out the loser. The proof is apparent: throughout the long
period of evidence gathering that started in 1927, then in testimony to the
IJC, and finally in the further investigations leading up to the Arbitral Tribunal hearings of 1937 and afterward until the dispute was settled in
1941, the two federal agencies operated at cross-purposes. Dean was a
consummate technologist who dazzled the Arbitral Tribunal with his expertise. He and his colleagues sensed they had the opening to replace the
Selby standard with a new industry standard, to be published in a USBM
bulletin once the Trail dispute was over. As he wrote a colleague, "I will
have in hand at the end of my work [for the tribunal] a very fine dissertation on the metallurgical phases of the case. It remains to be seen
whether ethics and policy with regard to American smelters will allow
the Bureau to publish it. In the first place it will be necessary to discredit
our Department of Agriculture very badly!"[62] Indeed, Dean was the
coauthor with Professor Swain—a consultant to C.M.&S. who was appointed by the Canadians as their adviser to the tribunal—of the USBM
bulletin on Trail. Their substantial report was published in 1944. None of
the voluminous data and reports by USDA researchers found their way
into the literature.

In contrast to the divided American effort, the Canadians were very
well organized. Sherley, a retired congressman, who had been touted to
Roosevelt as a vice presidential running mate, was a lawyer doing a job
he neither relished nor particularly believed in. He was persuaded by
Dean's argument in favor of technical measures and had little confidence
in USDA scientists whose results he could not understand. Sherley bungled the hearings. At Spokane, testimony by USDA scientists was deci-
mated by W. N. Tilly, the Consolidated lawyer, who punched holes in
their experimental methods and questioned the data from U.S. smoke
recorders, some of which had been dismantled for lack of funds, so the
readings were partly out of date. Canadian research reports were neatly
printed in multiple copies and contained new findings; U.S. reports were
typewritten—at Spokane as few as three copies of each was available—
and they mostly contained old field research findings. In fact, the U.S.
case was poorly prepared and badly executed, whereas the Canadian

researchers maintained without fear of cross-examination from Sherley that their experiments showed little damage to plants had occurred since 1933, and plants subjected to low levels of fumigation under controlled conditions could survive and then recover quite well.

As mentioned, in Spokane the settlers' power and influence collapsed in the progressively more desperate negotiations with C.M.&S. that went on simultaneously with the disastrous public hearings. Read, Dean, and a State Department lawyer all lent a hand in trying to arrange a settlement. From Trail, Blaylock was in close, sometimes hourly contact with Crowe in Spokane. This marked the end of the CPA's extraordinary and unusual influence in the Trail smelter case. Relying on Metzger and the USDA, the farmers, too, had hitched their star to the invisible injury thesis. Leaden had already suffered a second stroke in February, and as Raftis wrote to Metzger, who was himself dying, "I am fearful that he is going to pass out of the picture and that he will not live to see the end of this long drawn out fight."[63] Although the U.S. side did better in advancing the validity of its scientific evidence in the later hearings, with John Raftis managing the legal aspects, it failed to convince the three arbiters of the invisible injury theory. In the face of conflicting scientific testimony, the arbiters accepted only the evidence for visible injury and awarded $78,000 in damages for two burns causing visible markings in the spring of 1934 and 1936. Forty-eight farmers received $62,000 for cleared land and $16,000 for timberland.

Attention shifted to enhancing the dispersion regime as a way to reduce still further the transport of pollution into the United States. On the basis of subsequent experiments with wind currents and diffusion patterns, Dean and Swain recommended a more rigorous control regime that was adopted on an experimental basis for three years and then mandated by the tribunal, which disbanded in 1941.

This regime greatly reduced smoke visitations down the Columbia Valley and was considered an innovative step forward in the science of smoke dispersion. Used in conjunction with scrubbers, acid plants, and the other methods to treat and limit the production of SO_2, this type of regime employed intermittent, controlled reductions in furnace operations to limit smoke production when meteorological conditions warranted. In the industry it came to be called a Supplemental Control System, or SCS, and was a mainstay of industry until the 1980s, as will be discussed in the chapters on the Arizona smelters. However, had the theory of invisible injury been upheld, the regime would have been more rigorous, been more costly to enforce, and quite likely mandated production shutdowns in order to meet the higher standard. To control for invisible injury, a lower threshold of fumigations would have been required, and compensation for degraded farm- and timberlands would have been correspondingly larger.

The upshot was that the USDA abandoned the smelter fumes project; its reports were never published; and what had begun with high hopes for seeking redress for agricultural interests in general, and for Stevens County farmers in particular, while at the same time conducting innovative science and contributing to the literature on air pollution evaporated. Metzger's death was a devastating blow from which the U.S. legal case never recovered. Inadequate funding dogged their efforts after 1932, precisely when they needed to perfect and publish their data. Outmaneuvered by the Bureau of Mines, Knight felt that his team had been "sold down the river" by Sherley, aided and abetted by Dean and Swain. Furthermore, during much of the ten-year period he headed the U.S. Smelter Fumes Committee, "we were handicapped for funds and there was a lack of sympathy expressed in certain quarters for the [farmers'] group making and asking for redress."[64]

At best, the arbiters were disposed to sidestep the controversial findings of environmental damage. It would be decades before the science on the effects of low-level fumigations on plant life would be perfected and vindicated. And, in truth, it is doubtful if even today it could meet the proof of damage standard demanded by the law. To make appreciable headway, science and the law would have to be yoked together for effective public policy. In the Trail dispute, they were not.

Instead, the true significance of the Trail case is that it was an elaboration and perfection on the Ducktown case in two ways. Liability—the "polluter pays" doctrine—was extended to interstate pollution cases in the Supreme Court's ruling on the Ducktown dispute. The tribunal simply extended this precedent from U.S. law to transborder pollution cases. In addition, Ducktown solved its excess sulfur problem by creating a large artificial fertilizer industry that found a ready and profitable market in the tobacco and cotton fields of the South. Trail's Elephant brand fertilizer followed down this path as well.

The upshot is that smelter managers could resolve their pollution problems with a technical solution—flue recovery, acid plants, and a supplementary control system to minimize the effects of the SO_2 that still went up the stacks. In Dean and Swain's words, "a complete solution to the problem of damage to vegetation in the United States can be reached without an arbitrary limitation on the total amount of SO_2 emitted by the smelter."[65] If soils, forests, and crops could be shown to be subject to progressive degradation, such a limitation would not, of course, be arbitrary.

Understandably enough, Consolidated Mining and Smelting Company was determined to avoid a ruinous settlement, let alone a shutdown. Methods for recovering waste by-products were successful. But by successfully narrowing the proof of damage while simultaneously heading off a pollution standard that could have been applied to the entire smelting industry, it was reassured that if and when the polluter pays, it would not be very much.

How close were USDA scientists to a breakthrough? Or did the NRC scientists really have it right, as the ruling seemed to indicate? Assessing the promise of this science, the institutional and political obstacles placed upon this research, and the ultimate failure of the U.S. scientists to prove the invisible injury thesis is the subject of the next chapter. The continental dimensions of the dispute, something well known to the participants but then lost to memory, will be discussed in the concluding chapter on the Trail smelter case.

CHAPTER TWO

"This Most Illusive of Problems":
The Scientific Dispute

U.S. scientists had been carrying out field studies on crop and forest burns for only a few months in 1927 when the smelters became alarmed that these investigations could lead to more stringent emission standards being imposed on the entire industry by the courts. (The prevailing Selby Commission standard on permissible smoke emissions, with payments for damage caused by pollution, favored industry over agriculture.)[1] George C. Hedgcock, the U.S. Forest Service plant pathologist with many years of experience examining the effects of smelter pollution on forests and farms, reported a conversation with Trail's attorney (R. C. Crowe), who told him that "all of the other smelting companies are joining in bringing pressure on the Trail people to prevent the situation there from progressing to a point where the Government can officially call attention to it in a publication. They want at all costs to prevent the Govt. from broadcasting information that could be used against the smelting business generally." Trail was reportedly willing "to cut its SO_2 output in half and to construct by-products plants costing several million dollars," hoping that the investigations would then be called off.[2] As mentioned in Chapter 1, Trail management had already begun an internal study of the pollution problem in 1926, and a small experimental acid plant was being planned by 1928. But it was by no means certain that these efforts would be enough, because a possible result of the U.S. scientific investigation might lead to a far more stringent program of SO_2 recovery.

The bottom line was that very large sums were riding on the results of this litigation. The smelters wanted to stay in business and, if possible, turn a profit on unwanted by-products without (from their perspective) losing money on overly restrictive methods of smoke remediation. Their preferred option was to develop control technologies that would limit but not fully abate the air pollution problem. As already discussed, Trail was the industry leader in following this business strategy. The companies accepted the fact that large damage awards were part of the price of running a polluting industry. But they also wanted to avoid yearly, open-ended payments and to limit the size of these awards, hence the smelter lawyers' winning strategy of narrowing the definition of damage to actual, observable, economic damage while raising doubts about the scientific evidence on the basis of which broader injury could be claimed.

Consolidated carried the spear for an entire industry, the managers of which knew each other and acted as a group when fundamental interests were at stake. Read's correspondence contains several letters to American smelter executives at ASARCO, Phelps Dodge, and American Sulfur that show him to be fully conversant with the industry's perspective—for this was Trail's best defense—while tapping into a pool of expertise, shared experiences, and shared risks. (That the smelters had a continental reach and acted together will be treated at length in the next chapter.) Industry advocated a particular line of scientific thinking on the nature of this smoke damage, which defined the immediate ground rules for one side of the dispute.

From Salt Lake G. L. Oldright, a USBM official familiar with ASARCO's field experiments on the effects of SO_2 fumigations on plants, summarized what the smelting companies hoped to show:

(1) Solid toxic agents can be removed successfully in a properly designed plant.
(2) The theory of invisible damage is not tenable. . . .
(3) By properly controlling conditions, the SO_2 concentration in the atmosphere may be kept comparable to that obtaining in other industrial centers, and below the point at which the gas is harmful to vegetation, except under exceptional cases.
(4) When such cases arise the nature of the damage committed is such that it can be paid for without permanent damage to the property or to the State.[3]

What the smelters most wanted to finesse was "setting a rule as to the frequency, duration and amount of concentration permissible without causing damage or injury." This was the famous "yardstick question" that ASARCO researchers had been working on for years but were unable to pin down. After the First World War, Dr. P. J. O'Gara's "work there resulted in the conclusion that sulphur dioxide injury depended upon so many factors that no rule can be established under which smelters can successfully operate."[4] At least this was the industry's interpretation of the results obtained by O'Gara in Salt Lake and in experiments during 1917 at the Stanford University nursery in cooperation with Professor Robert E. Swain. Experiments in open field plots at Stanford failed to produce exact results: too many environmental factors were uncontrolled—in soils, atmospheric conditions, bacteria, fungi, or insect infestations—causing minor variations among plots. Having investigated the problem over two growing seasons at the nursery, Swain and his students concluded that "the probable error in this method is so high that the results on 'invisible injury' have not been conclusive." Too many minor variables intruded.[5] Thus the theory was not yet proved or disproved, but one way to attack it, by questioning research procedures, was established.

Indeed, O'Gara established strong arguments for sticking with the "visible injury," thesis, which is what the smelters wanted to hear. Subject to local conditions of temperature, humidity, and light, the threshold of visible leaf damage was ascertained by controlling for time and a given concentration.[6] According to this school of thought, the only form of plant injury that could be measured reliably was these external burns. The Selby standard was one part per million of SO_2 for three hours. Damage could occur at lesser concentrations, O'Gara found, and he circulated a list that indicated the relative resistance of a number of plants to sulfur dioxide. This list was eventually published in the 1950s, many years after his health broke and he stopped working in 1923.[7]

Few of O'Gara's research findings on the effects of very low level fumigations were ever published. It was not because he had found evidence the smelters did not want to hear—in fact, USDA scientists were granted access to his data—but because later improvements in fumigation cabinets made at the Salt Lake station provided more accurate measurements of gas effects at very low concentrations than the instruments available to O'Gara. Furthermore, his data on plant yield were considered untrustworthy because of the fluctuation between check plots. Nonetheless, ASARCO considered that "Dr. O'Gara's data on the immediate affects of fumigations of varying concentrations is particularly significant."[8] That is to say, O'Gara had been unable to pin down anything dangerous to the smelter position. Griffin thought the same way after visiting the station in Salt Lake: "I found that O'Gara had not done very much with these real low concentrations either. He had done a few experiments at concentrations of 0.3 to 0.35 p.p.m. with the time element at 50 to 70 hours, the results being inconclusive but mainly negative."[9] But much on their minds, then and later, was the possibility that O'Gara had missed something important.

Experiments on this problem were continued in Salt Lake by Dr. George R. Hill, who with the chemist M. D. Thomas designed and manufactured the automatic SO_2 recording devices later used by both the U.S. and Canadian scientists in the Northport area. They also improved upon the fumigation cabinets then in use. Hill and Thomas conducted extensive experiments on alfalfa, a plant that was particularly sensitive to smoke burns. Durwood F. Fisher, the USDA plant pathologist, reported after a 1929 visit he and Griffin made to Hill's Agricultural Research Station that while Hill had some data indicating cumulative effects of SO_2, . . . for the most part his results indicate that non-toxic or rather non-lethal doses are largely overcome by the plant and that thereafter it is capable of withstanding any dose up to the lethal one, provided normal conditions intervene between fumigations. This is information we really ought to have in order to answer the questions which have arisen in Northport.

Continued experiments with the new equipment showed not only that alfalfa could recover from low-level fumigations, but that neither the yield nor the food value declined in any cumulative way.[10] In his capacity as director of the industry's leading research station, Hill, a former dean at the Agricultural College of Utah before moving to ASARCO, appeared as an expert witness for Consolidated at both the IJC and the Arbitral Tribunal hearings.

Starting in 1925, meanwhile, in research at Stanford funded by ASARCO, Swain and his graduate students measured the effects of fumigations on wheat yields in a series of experiments in the Chemistry Department's basement. This research confirmed Swain's long-held belief that the invisible injury thesis was invalid. "The claim has often been made in smoke damage cases that the reproductive processes of plants are adversely affected by sulfur dioxide even if the leaf markings are absent," Swain and his graduate student Arthur Johnson concluded. In fact, "the large yield, well-filled heads, and plump well-formed grains of the treated plants point definitely to the contrary conclusion." Swain was a pioneer in the science of air pollution, and his career was one of great distinction. He served as acting president of Stanford University, was a founder of Stanford Research Institute and the Hoover Institution, and served as well as a frequent consultant on pollution cases for the U.S. government. In an age before air travel was common Swain was jokingly known as Stanford's "Pullman Professor of Chemistry" for his frequent absences from campus.[11] Swain also consulted for C.M.&S. and with the USBM's Reginald Dean served as technical adviser to the Arbitral Tribunal. From the USDA perspective, he was the most prominent U.S. academic in the "smelter camp," having been hired and funded over the years by industry.

The result was that industry's position was bolstered by strong scientific support. In a nutshell, as Fisher reported:

Smelting interests attach no significance to fumigations which do not result in toxic manifestations or definite markings of the foliage within a comparatively short period following exposure to the gas. They do not admit any cumulative effect of fumigations. The only injury they recognize is that which may be termed the typical acute burning of the foliage manifested in the killing of definitely delimited areas at the margin or between the veins of leaves. On conifers and often on other plants these areas are browned but on many crop plants and others they become bleached and the same pattern of injury appears on both sides of the leaf.[12]

Moreover, this line of argument bolstered a pollution control strategy based on recovering the by-products, developing additional markets for the new products (mainly fertilizers), and dispersing what gas remained. It was this overall agenda that the USDA men now challenged, excited as they were by the prospect of an interdisciplinary research effort and enticed by a different school of thought, mainly earlier German work, where the literature was sparse but suggestive.

Footnotes and citations in their research reports confirm their familiarity with the work of German chemists and plant scientists who, starting in the 1880s, had advanced the *unsichtbare Beschädingungen*, or invisible injury, thesis. Having been trained at Heidelberg and Yale, Swain himself was of course familiar with it. This early literature was summarized in Paul Sorauer's well-known *Handbuch der Pflanzenkrankheiten*, the third edition of which had been translated into English in 1922 under the title *Manual of Plant Diseases*. In this landmark text, the effects of heavy burns near a polluting plant that killed leaf tissue rapidly were distinguished from the effects of widely dispersed smoke that

> is usually breathed in by the plant slowly but permanently. The former effect, appearing rapidly and eating into the tissue, is distinguished as *acute* from the phenomenon of a slow poisoning which is termed *chronic injury from smoke*. Of course, the latter effect must have made itself felt inside the plant before the external characteristics appeared. The chlorophyll apparatus is changed . . . even if the plants still appear perfectly normal. In this case an *invisible injury from smoke* is spoken of.

Sorauer and his colleagues concluded that the invisible injury disturbance "could be averted very easily and the plant, as has been found, is in a position to cure itself after the cessation of a weaker fumigation."[13] For those seeking to abate smelter pollution, this argument led inexorably to the imposition of much tighter controls, and USDA scientists subscribed to it.

Important advances in the science of phytoxicity also were made by Hans Wislicenus at his respected Tharandt Academy. He was the first to specify the interplay of light and the opening of stomata, among other plant mechanisms, in the assimilation of SO_2 and subsequent plant damage. Wislicenus found that disturbances in photosynthesis were caused when the stomata were open during sunshine.[14] Active from the 1890s to the 1920s, his work was known to the Americans as well.

Invisible injury and the attempts to specify subtle changes in plant metabolism attributed to it first surfaced in the German literature in the 1890s. The theory was controversial until well after World War II, when the studies finally specified how low pollutant exposures affect vegetation either over long periods of time or permanently. In reviewing recent literature, two East German researchers concluded that "in the case of invisible emission exposure the measurable reactions of plant metabolism are mainly assimilation depression as well as changes in enzyme activities, in isoenzyme patterns and in further plant constituents." Only in the 1970s did it become clear that controlled, high-level fumigations conducted to develop thresholds of plant tolerance had little relevance to this other type of injury.[15] The importance of the former (acute injury) in providing proof of damage that could stand up in court will be readily apparent below. Proving the latter (the precise nature of a reduction of plant performance due to invisible injury), however, was exceedingly difficult.

In a comprehensive study of smoke damage to vegetation and factory emissions published in 1923, Julius Stoklasa, a Czech scientist, reaffirmed the three types of plant damage: acute, chronic, and invisible. Because it interfered with photosynthesis, causing a drop in crop yields, invisible damage deserved a great deal more attention, he suggested. He characterized the effects as the reduction of photosynthesis, accumulation of sulfate in the leaves, premature aging, growth depression, and increased susceptibility to secondary damage from insects, drought, and soil decline. Unavoidably, Stoklasa concluded, the efforts to reduce the damaging effects of smoke gas would lead to international negotiations, both to improve the health of the public and to strengthen the regulatory power of each nation-state for the common good.[16] Whatever the problems with air pollution being faced by the newly created state of Czechoslovakia—and they were serious—Stoklasa's message was controversial and not the one that North American industry wanted to hear.

Swain knew Stoklasa's book and discounted it. To him, the "invisible injury" claim, strongly supported by some of the German investigators and often made, had not been supported by convincing experimental evidence:

Much of the support for this theory has come from field observations made in the regions adjacent to plants where smelting and other industrial operations were in progress. The uncertainty of this evidence must be conceded. It is clouded by too many variables, [so that] clean-cut proof of [SO_2's] action as an agent of injury has often been lacking. The presence of tar and toxics other than SO_2 muddles the results. In controlled field experiments, one source of uncertainty is that minor variables in environmental factors . . . are difficult to control in open field work. Slight differences in soil, or in the local effects of rain or wind, may occur even among neighboring plots. Bacteria or fungi, or insect infestations, may appear in one plot and not in another. As a consequence the results of this method in the past . . . have often been inconclusive when it has been applied to the study of the effect of sulfur dioxide on plants at low concentrations below those at which acute foliar markings are produced.

And while the "extensive and painstaking" field experiments of O'Gara and later Hill deserved special mention, "their investigations . . . have been directed more to the causes and conditions and consequences of visible injury than to the question of 'invisible injury.'" To be sure, O'Gara had tried, and failed, to pin down the so-called "yardstick question." Swain and Johnson's own research on wheat, conducted under carefully controlled conditions, provided what amounted to very strong arguments against the thesis, as they maintained at the annual meeting of the American Chemical Society in September 1935.[17]

Swain was a gifted writer and presenter, and in discussing the complexities of research on SO_2 damage he spoke authoritatively, especially when litigation was involved. For years he had spoken out against the invisible injury thesis with its implications for tight controls. While serving as the court-appointed commissioner to oversee the resolution of two Utah smelter cases in 1920, he specified stack and furnace improvements on the ASARCO lead plant at Murray and imposed a control regime. His rationale for this decision explicitly rejected the invisible injury thesis. The earlier German research was done close to smelters and without the benefits of modern measurement devices, and, in fact, the theory had never been supported by any conclusive scientific evidence either because of the lack of an otherwise normal environment for both the test and the control plants or because of the lack of fumigation methods that were open to serious objections. These were essentially the same arguments that he would use fifteen years later in the coauthored wheat paper, indicating Swain's views were formed well before the Trail dispute began.

Early in the work of the Selby Smelter Commission, he continued, "a new era in the investigation of the effects of sulphur dioxide on plant life began." Using the new fumigation cabinets, it was shown that the effect on plants of very low concentrations of SO_2 was temporary, at best. All of the important recent work had been based on this equipment and the new techniques.

There is now every reason for believing that, at concentrations in air which are below those which will produce foliar markings, hence visible injury, sulphur dioxide is freely tolerated, and in all probability directly utilized by plants and that there are not effects, immediate or remote, on their general metabolism which lead to lowered vitality, to any interference with the processes of growth or production, or to any reduction in crop yield. When the limiting concentrations for non-injury under the prevailing conditions of humidity, light, temperature and duration of exposure are exceeded, and injury is done, this is primarily local, rather than systemic, and quickly leads to discolorations which are quite typical and pronounced. When this point is reached, and only until then, can it be claimed that injury has been done.[18]

With the authority of the U.S. court behind him, Swain's opinion left no doubt that the court should not base any sort of control regime on an unproven scientific theory. Given progress in fumes remediation, and the smelter's responsive effort to comply, it was open to question whether any regime at all should be imposed on ASARCO's Murray, Utah, plant (or the U.S. Smelting Company's plant at Midvale). Since so many interdependent factors were involved in plant operation, operating restrictions—for example, on the amount of sulfur eliminated—"should not be arbitrarily set up but worked out in conference with all interested parties."[19] In short, practical business considerations should be given due weight in questions of smelter fumes abatement.

As the Trail dispute progressed, it was very clear that all along there had been "two very different lines of thought in this case," as the attorney John Raftis, speaking now as the U.S. agent, told the tribunal in 1940:

One of them is the philosophy or the findings which have been built up first in Europe where they discovered this so-called subliminal or chronic type of injury arising from repeated doses or fumigations. We have tried in this country, principally by the American Smelting and Refining Company, to build up a theory that you cannot get damage unless you have markings; in other words you have to have a part per million to hit this stuff for about three hours, then you get a visible burn which you can go out and measure, as one of them says, as you would measure grain from a threshing machine.

He concluded: "They got by with this all right at Salt Lake, but that is not the problem up in northern Stevens County."[20]

Raftis then quoted at length from Swain's 1923 lecture on the lessons of Salt Lake, including his views on the invisible injury thesis, a position fixed long before, Raftis implied (to the judges' annoyance). At issue was the fact that, in direct contradiction to Griffin's findings, Swain and his former student Nelson had found no evidence of burn markings during their field survey of 1938. "How could this be?" Raftis asked when pressed by the judges about why views expressed in 1923 were relevant. "Yes, I will say that background and training and mental approach to this problem are very important in arriving at a decision. . . . Personally I think Dr. Swain is a brilliant and high type of man. But we all have different viewpoints in approaching problems, and I think that is his viewpoint; that is the viewpoint of the smelters in Salt Lake."[21]

Still at issue, after almost a decade of the testimony at both sets of hearings, was whether the effect of low concentrations of SO_2 on vegetation caused cumulative damage to plants, and, if so, what operating restrictions should be imposed to prevent it. This was no academic question, because Swain and R. S. Dean of the USBM were technical consultants to the tribunal, charged with devising and recommending a new operating regime for the Trail smelter.

The upshot was that the scientific battle lines between the two schools of analysis were clearly drawn well before the USDA went to Northport in the late 1920s. It remains now to decide on the basis of their reports and testimony what the American researchers found to substantiate the theory of invisible injury and, using the same internal evidence, to evaluate why the Canadian researchers found strong arguments for the visible injury thesis.

When Skinner and his colleagues returned from making a reconnaissance of the area in July 1928, they knew that "a difficult and complex research investigation" lay ahead.[22] They heard from farmers about the corrosion to metal fences, roofs, and screens that was caused by the smoke; how livestock were reportedly receiving less food value from hay and alfalfa; and they saw enough evidence of crop and forest loss to wonder if soils and water had been affected by lead deposition and the effects of SO_2. These findings all became part of the investigation.

The company's scientist, Roy Neidig, told them that vegetation was the best indicator of sulfur damage. His team had started out to do a very thorough chemical study of atmosphere, soils, and plants in the region but had abandoned much of this line of work, including systematic and comprehensive analyses of air samples that, "due to the rough terrain and especially the vagaries of air currents carrying the fumes, they soon found it impracticable to conduct."[23] These aspects of the work were precisely the ones that interested Skinner the most. As he wrote Griffin, installed in Northport since the fall, he wanted to get the air volumes study (atmospheric sampling for SO_2) "out of the way so we can take up as originally planned the foliage work early next fall or winter." He also anticipated finishing up the soil work and the USBM's fume control study so that more of the budget would be freed up "to the strictly chemical work" in the next fiscal year.[24]

When the Skinner party was taken through the plant by R. C. Crowe, the company attorney, Trail managers "stress[ed] the claim that there is no 'invisible' injury from fumes. They quote unpublished work by Swain, of Stanford University, but this is at variance with earlier work by Sorauer in Germany," Fisher commented in his report, and in so doing he showed just how much this scientific controversy was on their minds.

With the exception of George Hedgcock, who worked on forests, Skinner and his team from the Department of Agriculture had little or no prior experience with the SO_2 problem. Skinner himself was a pure water expert and had not published in the field of air pollution. Griffin was coming from a long assignment as a chemist at a government arsenal (where he almost died from arsenic poisoning). Fisher was an orchard specialist, who since 1913 had headed the U.S. Fruit Disease Field Laboratory in Wenatchee, Washington, for the Bureau of Plant Industry. As man on the scene in Wenatchee, Fisher was given responsibility directing the plant study side of the smoke investigation. (In none of the voluminous files on the smelter fumes investigation with respect to chemical research on plants did I find references to prior scientific work [forestry excepted] either at USDA headquarters in Washington, D.C., or in the field.) To be sure, the USDA had made studies of the beneficial effects of sulfur on the growing cycle. Given their long-standing interest in the effects of smelter pollution on the national forests, however, it is certain that the scientific merits of the Selby standard were the subject of debate, and Skinner had at least one strong argument with a colleague in the Bureau of Standards who upheld it.[25]

In the summer of 1928, while George Hedgcock began his timber survey, other members of the USDA studied soil conditions and surveyed the economic damage caused by the fumes; Griffin did the air sampling and managed the chemical side of the investigation from Northport; and Fisher, assisted by M. C. Goldsworthy, another plant pathologist, conducted fumigation experiments on apple trees at Wenatchee. It was a cooperative effort by the Bureau of Chemistry and Soils, to which Griffin reported, and the Bureau of Plant Injury, employing Fisher and Goldsworthy; both bureaus had laboratories at Wenatchee.

The most important plant work was done from 1929 to 1933, and initial results were encouraging, as Fisher reported to Kellerman, the bureau chief: "Our preliminary work indicates that there is a chronic type of injury to plants from repeated fumigations with low concentrations of SO_2. This, of course, is not accepted by the smelting interests and there is very little corroborative evidence in the literature." To him, the most significant results of the investigation so far were twofold:

Our experiments indicate that the effects of SO_2 fumigation, aside from the *acute* injury which is generally recognized, are to intensify drouth reactions. In a non-irrigated section such as that in the North-port region, this is a very important point and supports the con-tention of the farmers of that region. We also have secured evidence in these fumigation experiments pointing to a lowered food value of alfalfa subjected to fumigation and exhibiting what we term *chronic* injury.

The results of all this work would have a fundamental bearing on the case, and for Fisher and his colleagues, "we think it is important enough to justify considerable additional work." But hearing that a formal publi-cation had been scheduled by the Smelter Fumes Committee for spring 1930, he advised against it until this experimental work was completed.[26]

Griffin did not underestimate the difficulties. In writing to his boss, Skinner, after the visit to ASARCO, he thought that "it is going to be dif-ficult to substantiate our claim that these relatively low concentrations are so damaging to plant life, as there is very little literature pertaining to these low concentrations."[27] They knew the German literature and had O'Gara's field notes on the threshold of injury on various plants, out of date though they now were. So far, these data had only been interpreted by the smelters. "The index prepared from this work and the conclusions that may be drawn therefrom at present constitute the only yardstick available" to determine the relative tolerance or resistance of plants to SO_2.[28] Like Thomas, they were finding flaws in O'Gara's work.

But after visiting Hill's operation, Griffin realized that their own measuring devices needed upgrading to match the Salt Lake equipment. There the fumigation cabinets were larger than the one he had built at Wenatchee and included a "measurement of light intensity, a control of humidity conditions within the cabinet, and a metering out as well as into the cabinet of the air sulfur dioxide mixture . . . [which means] it is pos-sible to reach an approximation of the amount of sulfur dioxide actually absorbed by the plants themselves." They could not finish up after only two seasons, not if Griffin was to construct a larger fumigation cabinet, while Fisher, in response to what he saw at Salt Lake, set out some new plats in which instead of broadcasting or drilling, the alfalfa and grain were set out carefully in rows with strips of celluloid covering the ground in between.[29] The equipment was duly modified by August, and George Hedgcock used it for experiments on young conifers, while Fisher and Goldsworthy conducted a new series of experiments on alfalfa, wheat, and gaura, a weed plant very susceptible to injury by sulfur dioxide. The chemical phase of the project was now in high gear.

The plan of work was threefold: air sampling, plant experiments, and an areawide sampling of leaf tissue for sulfur content. In all three areas exciting new results were reported to the Smelter Fumes Committee. The results of 250 field trips in 1928 and 1929 showed that fumigations of more than 1 ppm were rare, nor did the higher concentrations occur for any considerable length of time. Rather, pollution of lesser intensities affected some 100,000 acres of U.S. territory down to Kettle Falls and correlated nicely with observed damage to forests and vegetation (see Hedgcock map in Chapter 1). In July 1929, SO_2 was found in the air in the general vicinity of Northport on an average of about two out of three days, or about twenty days a month. Readings for July 1930 were consistent with these earlier data on the portable recorders Griffin used. (Later that year, two automatic recorders, purchased from ASARCO, were in operation near Northport and Evans. The NRC team from Canada also installed two of these machines.) Given the prevailing wind patterns, dry soil conditions, and the unusual channeling of the fumes down the river canyon, readings clustering in the .3–.4 ppm range did not cause acute injury to plants, the only standard accepted by the smelters, but injury was observed nonetheless. And "as the literature provides no satisfactory solution of the question, it was necessary," in Griffin's words, "to undertake the fumigating experiments in order that the facts might be made known."[30]

At Wenatchee, Fisher and Goldsworthy set out to prove that such dilute and varied concentrations could cause cumulative and chronic injury to vegetation. Fumigations with atmospheres established synthetically that were similar in intensity and duration to those prevailing at Northport were begun, under comparable conditions of temperature and humidity. Thanks to these experiments, Griffin concluded, "there is no longer room for doubt but that appreciable [i.e., chronic] injury to various species of plant life may be expected when such plants are repeatedly subjected to the influence of low concentrations of sulfur dioxide in the atmosphere."[31] To appreciate what they were coming up with, it is well worth quoting at length from the experiments on alfalfa. They experimented first on fruit trees (Fisher's specialty), then on alfalfa, barley, oats, and various grasses using tents and a fumigation cabinet that had no means for humidifying the air, a serious shortcoming that was rectified in the newer cabinets Griffin installed after the visit to Salt Lake. The Stevens County smoke zone was somewhat drier than Wenatchee, which averaged two to three more inches of rain a year. The second set of experiments began in August 1929, and this time, Fisher reported, "all the factors affecting the results were under control."

In the tests for acute and chronic injury, alfalfa leaves in one test plot, fumigated for three hours at 1.33 ppm, exhibited fading on the upper surfaces, then both sides became involved, while other leaves took on a chlorotic appearance indicating a gradation effect from SO_2, producing acute injury. Further fumigation appeared not to enlarge lesions on individual leaves, but the grass became reddish-brown at the tips, then purplish in color as growth appeared to stop. A comparison of the test plot with check plots detected only slight differences, but "it was found that the injured fumigated alfalfa had a lower percent of total dry matter and of insoluble nitrogen (including protein content, the chief food value in hay) but an increased amount of total carbohydrates." In another test plot, fumigated for three hours times at 0.62 ppm,

> the first symptoms of chronic injury were noted on alfalfa on August 21. This was exhibited by the browning of the leaves and the purpling of the under side of the leaflets. This was particularly noticeable on the terminal leaflets the effects bearing a very marked resemblance to those produced by drouth and commonly found in the Northport area. Checks remained green and normal in appearance.

In a third test plot, fumigated twice at 0.42 ppm, "a slight chlorosis was observed on some of the leaves and this became intensified and extended to other leaves following the 2nd. fumigation." No further increase of typical (acute) injury appeared when the plants were left alone, but symptoms of chronic injury did appear.

Still another test plot, fumigated three times for four hours at 0.38 ppm for three successive days, exhibited symptoms similar to test plot three. Chemical analysis of the test and check plots

> showed very striking and significant differences. It was found that the fumigated alfalfa had a higher percent of dry matter, supporting the field observation of the more succulent condition of the check plot. It was also found that the fumigated alfalfa had a much higher percent of total carbohydrates and a markedly lower percent of insoluble nitrogen. This indicates a lower food value of the fumigated material. In general the results of analysis can be interpreted as indicating that the effects of fumigation are exhibited in a more advanced maturity or premature ageing of the shoots.

The results in terms of food loss in this plot tested for chronic injury were similar but even more pronounced than in the first test plot, which it will be recalled was subjected to a much higher concentration of the gas, producing acute injury. Another test plot of barley, fumigated at 1.48 ppm, exhibited tissue collapse after two and one-half hours.

From this series of tests, completed in August 1929, they concluded it was now proved that typical (acute) injury could result from the cumulative action of repeated nonlethal fumigations with low concentrations of SO_2. Also, according to Fisher, "markings of a chlorotic condition which generally preceded actual collapse and killing of limited areas indicated the first effects of SO_2 on chlorophyll." While the limit of tolerance had not yet been exceeded (i.e., where acute injury was produced), the gradual intensification of chlorosis "indicates direct toxic effects on the chloroplasts even though actual killing of tissue did not result." Chronic injury "operates both in starving the plant through the effects of chlorosis and in depriving it of necessary water holding capacity through the intensified drouth effects produced." In turn, the signs of chlorosis also indicated a disturbance of nutritive processes, lowering food value.

Fisher's overall conclusion was bold:

> The results of these experiments can be interpreted in no other way than to prove that there is a cumulative toxic effect of SO_2 on plant tissue; that this effect is exerted particularly on the chloroplasts, and that it may result either in death of the most severely affected structures, as in the usually recognized SO_2 markings, or in a persistent injury exhibited as chlorosis.[32]

Here, it seemed, was both proof and confirmation of a main tenet from the German school.

A third aspect of the chemical research dealt directly with the invisible injury thesis, which was tested by sampling leaf and needle tissue from the smoke area for traces of accumulated sulfur. As expected, samples of foliage taken at different points along the Columbia River showed that the amount of sulfur accumulations were a direct function of distance from the plant—low in Kettle Falls, higher in Northport, highest at Trail. The species collected by the plant specialists included foliage from quaking aspen, apple, maple, buckbrush, birch, hazel, ninebark, chokecherry, western larch, yellow pine, Douglas fir, lodgepole pine, and juniper. Laboratory analysis of injured and normal leaves from the same locality revealed a strange anomaly: "In about one-half the total number of instances the sulfur in apparently normal leaves is greater than that in injured foliage, and in the other half vice versa."[33] Why, then, did certain plants die? According to Fisher, "The only conclusion that can be drawn is that the living material has just about reached its limit of tolerance for SO_2 and that the addition of even a slight amount, perhaps in some cases an infinitesimal amount, would be more than the tissues could stand and they would thereupon succumb."[34]

In other words, leaves reaching their saturation point had been suffering with the effects of invisible injury. "In cases where no injury will ever be readily apparent to the eye, growth may be retarded by chronic invisible injury." Trees in stressed environments were the most vulnerable. Well-watered, fertilized trees such as those found in the Stroh orchard maintained by C.M.&S. could better withstand heavier accumulations of SO_2 than leaves from trees of low vitality that were just struggling for existence. Some leaves had a higher tolerance for sulfur than others from the same plant; some healthy leaves had a higher sulfur content than the injured leaves. L. R. Leinbach, the chemist doing the laboratory analysis back in Washington, concluded:

> While at first glance it may seem anomalous that apparently normal levels could ever have more sulfur than those showing visible injury, the data seem to show either one of two things, or both:
> (1) There is a variation in the ability to absorb SO_2 among the leaves of a plant, due to some factor in the plant itself;
> (2) Some leaves show injury before others, with the same sulfur content, show it, due to differences in exposure to sunlight, air and other environmental influences.
> Once a leaf is badly injured or dead its power to absorb SO_2 is evidently very materially lessened. A high sulfur content, without apparent appreciable injury, simply represents a state of unstable equilibrium; the slightest additional influx of SO_2 or the slightest unfavorable environmental influence will be apt to lead promptly to the death of such a leaf and hence eventually to severe injury or death of the plant.[35]

That final increment of sulfur dioxide is what proves to be the straw that breaks the camel's back, Griffin summarized. It "may be present in the atmosphere in very low concentration, the progressive, accumulative effects of preceding weeks or months providing the conditions necessary to the making of the last sulfur dioxide visitation visibly effective. *In this sense the phenomena of so-called 'invisible injury' to plant life becomes a demonstrated fact.*"[36]

It bears repeating that the chemical research took pride of place in this initial stage of the U.S. investigation. Skinner and the Smelter Fumes Committee also authorized soil work, which at first did not produce results in which they had confidence, the Hedgcock forest survey, which went well, and an analysis of economic damage. This chemical evidence was the keystone of the American case against the smelter, which was offering only to reduce pollution by one-third and with gradual reductions thereafter. "Since it has been established that damage [from sulfur dioxide poisoning] is cumulative," the committee said, even a reduction of this amount would still cause damage to Hedgcock's first two zones, and without the reduction all zones would continue to enlarge.[37]

Victory was in the air. The Americans' decision to reject the IJC ruling rested on this evidence—a decision stiffened, as we have seen, by pressure from the settlers. The research had gone well; Fisher was promoted and sent to Washington, leaving Goldsworthy to finish up at Wenatchee, while Griffin expected reassignment to the West Coast.

Meanwhile, the Canadians had been conducting surveys of their own. It was the threat posed to the smelter by the possible results of U.S. inquiry that had prompted Ottawa to send its own scientists to the field "in order that all the facts relating to Canadian interests in this dispute would be made available for consideration by the International Joint Commission."[38]

Meeting for the first time on September 10, 1929, the Associate Committee on Trail Smelter Smoke assigned Dr. George H. Duff, a senior plant pathologist at the NRC, to head a team of some eight researchers and administer the field research program. The scale of this effort bespoke the urgency of H. M. Tory, head of the NRC, and his colleagues to check on the U.S. evidence and, given sufficient scientific grounds, to counter it, especially since Consolidated wanted an NRC team in Northport to establish a credible, independent research effort. Soon the core team was in place: Duff, assisted by Dr. Morris Katz, who as the person in charge of the chemical phases of the investigation was Griffin's counterpart; A. W. McCallum of the Department of Agriculture, a forest pathologist and thus Hedgcock's counterpart; and George A. Ledingham, an assistant plant pathologist, who paralleled Fisher. F. E. Lathe, the project officer in Ottawa, was Skinner's counterpart on the U.S. Smelter Smoke Committee. Thus a full-range parallel investigation was launched, including chemical aspects, soils, timber, meteorological conditions, and an economic survey but excluding a survey of health aspects, because the Americans were not doing it.

It was Katz who installed the first automatic recorder, at the Stroh experimental farm, prompting Griffin to locate another of these ASARCO-made machines in Northport in order to check on the Canadian's results. Although personally reluctant to undertake the additional burden of servicing these machines, Griffin complied at the State Department's request. In short order a second set of stations was installed at Marcus and Evans, respectively.[39] Prompted by the accurate readings of low-level fumigations they were getting from these sensitive machines, the two chemists compared notes. "I had quite a talk with Katz," Griffin reported in May 1930. "He told me that for some time now he has noticed that, when his machine records $\frac{1}{2}$ part of SO_2 per million parts of air for a number of hours, he can generally detect SO_2 markings on the alfalfa." Based on this observation, "he stated that he was convinced that Fisher and I were 'on the right track' when we told the Commission that an SO_2 concentration of 0.4 to 0.5 p.p.m. applied on alfalfa for a total period of 12 hours (4 hours per day for 3 consecutive days) would injure alfalfa." As Griffin told his boss, Bill Skinner, back in Washington, "This simply confirms my opinion that the key to the situation rests in our fumigation experiments," to which Skinner replied: "It is of course pleasing to know that he [Katz] has confirmed your observations."[40] In any event, a prolonged scientific duel was about to begin.

It happened that on most of the questions relating to the cause of injury, witnesses before the tribunal, while "completely honest and sincere in their views," expressed contrary views and arrived at opposite conclusions. "Unfortunately," the tribunal was to conclude, "such field observations were not made continuously in any crop season or in all parts of the area of probable damage; and, even more unfortunately, they were not made simultaneously by the experts for the two countries, who acted separately and without comparing their conclusions with each other contemporaneously."[41]

If the Americans showed little enthusiasm for joint research, the Canadians needed little prompting to embark on a sustained effort to duplicate and discredit the American field research. The uses to which their evidence was put by the legal staff at Consolidated was laid out in considerable detail in a two-part volume entitled *Trail Smelter Question*, which was published in 1939 and distributed on a confidential basis "to those smelters, industries and other parties who were interested in the outcome of the case because certain principles of sulphur dioxide injury were being advanced which, if sustained, would have a material bearing upon sulphur dioxide nuisance litigation generally."[42] With respect to the scientific contest, this publication is, in short, a victory brief for and by industry. The actual progress of their research is detailed in reports and minutes of the NRC's Associate Committee on Smelter Fumes Investigation.

In August 1930 the Americans invited McCallum, Katz, and Duff to observe their fumigation experiments at Wenatchee. Far from being convinced by what they saw of Goldsworthy's experimental plots, the Canadian field staff returned

unanimous and decided in its opinion that we cannot maintain an independent opinion on the questions of what are and what are not forms of SO_2 injury and whether there is such a thing as an innocuous amount of SO_2 for crop and forest vegetation, unless we have the opportunity for independent investigation. Nor can we accept Dr. Goldsworthy's position from what we have seen of their experiments at Wenatchee.[43]

Duff noted, first of all, that insects were very numerous in both the test and control plots (cages were being prepared to prevent insect injury). Second, the Canadians felt that the gradual appearance of symptoms of injury they saw was at least in part a result of siting the plots in an orchard where very high night humidities obtained, giving a stimulus to the action of SO_2 on vegetation. In short, the intrusive effects of uncontrolled variables, long identified as a substantial problem in open field experiments, now brought *this* research into question. Thinking that the USDA results might not stand up, they proceeded to duplicate the tests in parallel plots at the Dominion Experimental Farm at Summerland, British Columbia, and were careful to conduct a second set of experiments at Stroh, under local conditions.

Concerning the American experiments themselves, Duff wrote that the so-called signs of cumulative and progressive injury, as indicated by the slow development of symptoms, were likely caused by climatic conditions. He also questioned the claim that invisible injury to leaf and flower buds was occurring: "Experiments upon a very sensitive plant [like Gaura] which must involve almost complete loss of foliage, cannot be taken to give a picture of the direct effect of SO_2 upon flowers." That injury was sustained at low concentrations was "clearly proved," he said, "but the injury is nominal only and in every practical sense negligible. The appearance of slight classical symptoms of injury is taken by Dr. Goldsworthy, however, to imply a succession of subsequent secondary forms of injury which are serious. This critical point in the argument was not supported by what we saw. It is such an extremely important point, however, that in our view it demands independent investigation." And finally, he stated, it will be claimed that "concentrations which do not produce the classical symptoms of injury nevertheless produce the subsequent secondary types of injury."

The Americans were explicit about the conclusions drawn from this research. In Duff's words, "Dr. Goldsworthy told us definitely that they held the view that SO_2 was injurious in any concentration however low. It is evident that they are preparing to defend the position that no gas whatever from Trail can be permitted to pass the boundary."

This was not precisely the American position, being more a case of Goldsworthy aiming for the perfect at the expense of the good. Nobody was arguing for complete abatement; rather, in light of the scientific findings, how much would be required? With the IJC preferring a technical solution rather than outright abatement, the U.S. Smelter Fumes Committee early the next year phrased it another way: "The concentration of sulfur dioxide at the international boundary should at no time exceed a maximum fixed by results of research and such maximum of concentration should be of short duration and of infrequent occurrence."[44] In effect, what this research was showing was that the smelter would need to adopt a very strict operating regime to stay in business.

Bearish though he was on the American experiments, Duff was sufficiently impressed to recommend that the committee give this matter its urgent attention, "involving, as it does, the whole question of the terms in which the company may be restricted in the matter of future discharge of gas over U.S. territory." The Smelter Fumes Committee—with both Crowe and Read attending—agreed to begin experiments at once at Summerland, "on an adequate scale, and with all possible precautions in order that there could be no possibility of questioning the results."[45] Katz was put in charge. And so, as the Americans prepared to publish their reports on the basis of two full years of field experiments, the Canadians belatedly, but expeditiously, got into high gear.

Griffin secured a copy of the Duff report for Goldsworthy, who noted that the NRC men had not seen the information he had accumulated on the loss of moisture by the leaves under controlled conditions. But "Goldy" did think it might be necessary to conduct experiments in a non-fumes area adjacent to Northport, which would be less convenient but better for experimenting on grains and conifers under local conditions. Furthermore, Metzger and Murdock, the State Department attorneys in charge, wanted this study done. Fisher, now installed in Washington, D.C., as chief horticulturalist at the Bureau of Plant Industry, was confident in the results: "I do not see where there is any loophole left in the fumigating work." But he, too, was "apprehensive over a possible question arising out of the location of the work" so far away from the smoke area. Results from fumigation experiments near Northport "as a whole might be strengthened, so far as controversial features of this case are concerned."[46] Skinner agreed. This same questioning of experiments conducted in conditions dissimilar to those existing in the Northport area was the tactic used by Metzger at the IJC hearings in arguing against accepting Hill's work on alfalfa in Salt Lake. In December, the Smelter Fumes Committee agreed to go ahead with check fumigations, perhaps in the valley below Marcus, near Kettle Falls.[47]

The Canadian Fumes Committee met again for the fourth time in Ottawa on February 14, 1931. Research on the sulfur content of leaves showed a decrease as one went down the valley, to slightly above normal at Kettle Falls. Soil acidity was not significant, requiring no further work. Reports on crop losses and forest conditions were also given. Although it appeared the damage to crops was greater in 1930 than the year before, the growing season was much better. Interestingly, the Canadians found that reproduction in the forests was taking place right up to the border, where conditions were favorable, but that insect depredations were an important factor in the destruction of timber. Possibly the most important work of the 1930 season had been examining Hedg-cock's test plots, where "the trees in general appeared to be in better condition than might be expected from [his] reports." Fumigation experiments on conifers took place at Stroh, and the work program for Katz, McCallum, and another chemist at Summerland was approved. In particular, the effect of increased sulphur content on the protein and carbohydrate content of barley would be investigated in relation to invisible and economic injury.[48]

At summer's end, conclusive evidence against invisible injury had not yet been obtained, but Katz obviously expected to get it during further testing of plots with a gas history comparable to crop plants in the Northport area. This would be done by duplicating the SO_2 readings obtained from the automatic recorder at Stroh farm. In the one long fumigation program on barley plots to test for the effect of SO_2 under varying conditions of light, temperature, and humidity to the point of injury, "a number of plots happened to escape injury altogether and were watched with interest subsequently. The sulphur content of the treated plants was increased but the protein content remained normal. The subsequent growth appeared normal." One test series did not constitute conclusive proof against invisible injury, an important objective since "claims have been made that these crops are invisibly injured and their quality deteriorated." But the research on 125 plots of barley and alfalfa and 43 tree plots was now getting into high gear at Summerland.[49]

Meanwhile, the IJC had issued its report in February 1931, laying out a large award to the farmers while validating the technical and economic approach to fumes abatement being advocated by the company. Nonetheless, the U.S. scientific team felt vindicated and prepared to publish, while the Canadians conceded nothing and accelerated their research. At this point, the Trail dispute went into diplomatic limbo, and the Great Depression was at its height. Fieldwork virtually ceased, although both sides maintained their automatic recorders in order to monitor emissions from the plant. The case came to life again in early 1933 when the outgoing Hoover administration rejected the IJC settlement, leading to the appointment of a three-judge Arbitral Tribunal and a recrudescence of scientific activity, this time favorable to the Canadians. In hindsight, it seems clear that the U.S. scientific effort never recovered from a loss of momentum in 1933, and budgetary data provide unambiguous proof, as Table 1 reveals.

Initially, the Americans outspent the Canadians two to one. Starting in 1932, however, the yearly congressional appropriations to the State Department for this purpose were much harder to secure and were reduced to boot. By contrast, Canadian funding (provided by Consolidated) more than kept pace. When the dispute came up for the second set of hearings in 1937, the USDA experts had to make do with experimental data com-

Table 1. Expenditures on the Two Smelter Fumes Investigations

Year	USDA	NRC-Canada
1927		$83,307.84[2]
1928	$35,000[1]	[1927–1931]
1929	$40,000	
1930	$40,000	
1931	$40,000	
1932	$40,000	
1933	$20,000	
1934	$19,000	
1935	$17,555	
1936	$15,000	
1937	$15,000	
1938	$15,000	
1939		
1940		
	$296,555	$227,980.70

Notes: 1. Funds transferred from the State Department, in US$. 2. Canadian government funds, in C$. 3. Consolidated Mining and Smelting Company transferred to the NRC, which asked for an additional 10 percent overhead on this sponsored research. Sources: Memorandum, McHarlow to Hackworth, July 1, 1940, U.S. National Archives, RG 59, 711.4215/1460, box 2175; memorandum, J. G. Read to the prime minister, October 27, 1937, in Canadian National Archives, RG 25, file 103-XI, volume 1691.

ing mostly from the early 1930s, after which their fumigation cabinets were dismantled and stored away. Their field data were similarly out of date and mostly reflected damage occurring *before* the acid plants and chemical recovery methods at Trail had come on line. Again by contrast, the adequately funded Canadian team produced up-to-date results that reflected the effects *after* the abatement measures began to take effect, and their equipment was deployed to produce the most current experimental results on plant injury. The result was that the Americans, their research having been front-loaded, started the debate, but the Canadians, having been sustained as their work progressed, finished the debate.

U.S. funding was inadequate from the start, and even the first installment of $35,000 in 1928 had been questioned: could it be cut to $15,000? A USDA staffer replied that

it would be better to do nothing than to attempt to do anything with $15,000. The problem is terrifically complicated at best. $40,000 for three years is entirely inadequate. I don't mean to say that some things could not be done, but what could be done would not be worth the doing. The opposing parties, the smelters, in the past 15 years have spent over a million dollars and have employed experts and concealed all their results, and in comparison what we will be able to do with $40,000 would cut a rather ridiculous figure. We ought to have $50,000 a year for 20 years—that would be the ideal thing—and would amount to about what the smelters have spent.

Actual expenditures for field research were much lower, starting at $10,897 in 1932 and declining steadily to a low of $3,767 in 1938. "These figures," Knight reported to the Smelter Fumes Committee, "are perilously low."[50] They were precarious as well because the USDA personnel had to live and work with a high level of uncertainty year to year about the actual fate of the appropriations, which both diminished the research effort and threw it off pace.

By contrast, the NRC research team could depend on financial backing from Consolidated, starting in 1931, and in addition the smelters actively cooperated by sharing advice and research results with them as the urgency to find countervailing evidence picked up in the mid-1930s.

The first incidence of company reimbursement for NRC research occurred in 1930, when the Russian-speaking Doukhobors protested loudly against smoke damage, which had been quite severe that year, to their farms in the Champion Creek area. When aroused, the Doukhobors, living apart from the Trail community that they served as carpenters, fruit vendors, and household help, manifested their hostility to secular authority in often bizarre ways, including public nudity. They were a long-standing annoyance, and Blaylock treated them with care. Accordingly, Dr. Duff was called in to estimate the damage, and Crowe picked up the tab ($3,400) and the cost of the Doukhobor investigation.[51]

Unaided, the NRC with its smaller civil service pool could not hope to continue matching a research effort the size of the Americans', and for its part the company felt justified in calling on the government to help, because private claims were being supported by the U.S. government. But could impartiality be maintained? Yes, if the Fumes Committee, rather than Crowe, had full authority to set the scientific program. Even more to the point in preparing for this second phase of hearings, could the *appearance* of impartiality be maintained? Yes, if periodic statements of expenditures, rather than detailed expense vouchers, were submitted to Crowe for approval and payment. To be avoided, Read was informed by the NRC treasurer, were grounds "for any suggestion that the C.M.&S. Company had any control in these investigations other than to reimburse the NRC for expenditures incurred on projects which the committee in charge might sanction."[52]

Alarmed, the normally phlegmatic John Read reacted momentarily like the combative and engagé Jacob Metzger, his U.S. counterpart. From the perspective of upcoming hearings before an international tribunal, "[this] investigation is rapidly ceasing to be a study, conducted by the NRC, under directions from the Government, and is becoming an investigation conducted by the Company, using scientists loaned to the Company by the Council." Anything to weaken the legal case would be unfortunate, General McNaughton, the incoming NRC president, wrote to Warren: "An unfavorable impression of this kind would also be very unfair to your company, since my research officers inform me that you and your officials have always shown a most commendable desire to bring out all the facts and to conceal nothing which might adversely affect your case."[53] Consolidated had been providing support for the fumes research since 1931, and now that a considerably expanded research program was under way, both the NRC and the company needed to reach agreement on accounting and reporting measures.

Read *was* worried about going over the line, and by the standards of today the Canadians surely did. Consolidated paid for publishing the NRC papers in a hardbound volume entitled *Effect of Sulfur Dioxide on Vegetation*, and Crowe asked to see the manuscript before publication; in gratitude he provided honoraria to government scientists when the final outcome was known and leaned on the NRC to pass on its Summerland findings on alfalfa to Hill in Salt Lake. Crowe's summary of the Canadian case and supporting evidence, printed in hardcover as the *Trail Smelter Question*, was distributed on a confidential basis to industry. By contrast, the State Department's control over the USDA research was direct.[54] To be sure, this control was done with publicly appropriated funds for government research. But this is a distinction without a difference. In the polarized climate then prevailing, both sides had so blurred the line between scientific evidence and legal advocacy as to cross the point of no return.

In contrast to the beefed-up Canadian program, the USDA limped along from appropriation to appropriation, barely able to sustain three field recorders and a modest follow-up research effort. Having been reassigned, Griffin and Fisher were put back on the case with inadequate staff support. What began in 1927 with high hopes became, after years of declining support, a fiasco; there is no other word for it.

To end the investigation would be unfortunate, Knight testified before the House Appropriations Subcommittee in 1932:

We feel these investigations must be continued until the case is settled, because the other side is carrying on investigational work and every year new facts are developed. . . . The tendency is for the investigation to become more extensive. In fact, we are finding difficulty in carrying on all the lines of work that we feel ought to be carried on with the funds allocated.[55]

Nevertheless, the House disallowed his fumes research budget for FY 1933 on the grounds that testimony had already been given in the case and while a decision by the two governments on the terms of settlement was pending. Being part of the State Department authorization bill did not help, because State, with few domestic constituents, was a traditional target for budget cutters. Knight asked the Washington congressional delegation (Representative Hill and Senator Dill) to get active.[56] The Senate then restored half of the funding, but the cost of running the three recorders alone was $15,000, and a further $10,000 was needed just for the chemical research.

With funding cut to the bone, maintaining one recorder at Northport was top priority, followed by the economic studies and Hedgcock's forest damage studies. Knight felt that they

> would incur a very serious loss if we did not continue our fumigation investigations . . . [at Wenatchee, but] on the other hand we had sufficient information at the present time to shake the evidence in the Selby Report. . . . While further work on fumigation would be highly desirable it was not as important from the legal standpoint as to follow the smelter smoke fumigation at Northport.[57]

Commenting on what to ask for in FY 1934, Metzger recommended that the work could be done for $20,000. However, "failure to provide money to continue the investigation would be abandonment of several hundred residents of the State of Washington and acquiescence in the devastation of [the land]. Such a course on the part of the Government would . . . be cruelty in all the circumstances."[58]

In 1933, the House Appropriations Committee was informed that the Marcus recorder was now shut down because of funding cuts, and "unless this government intends to abandon these western farmers to their sulphurous fate, these records must be maintained in their complete form." On this reduced budget, Knight added, "we are collecting certain material which we will not be able to compile." Nor were the written reports being kept up to date. The real point, a skeptical congressman concluded, was that a great deal of money was being spent to establish a claim for two thousand farmers.[59] Although that number was high—the real figure was around five hundred—it was hard to gainsay the fact that a powerful smelter was pushing American nationals around. But what to do, with fumigations continuing, if not continue at least a minimal investigation?

Skinner made the point again in pleading for a budget in FY 1935, but was it was almost a no go:

> There has been organized and trained for this special work a small but exceptionally expert personnel. There has been designed, constructed and installed at strategic points special types of unusual

equipment. All of this has been done at a very considerable expenditure of public funds. To interrupt the continuity of the work and to disperse the highly trained staff will be most inefficient and uneconomical if the work is ever to be completed and final adjustment of the dispute secured.

Congressman William B. Oliver of Alabama was reported to be vigorously opposed to any continuation of the appropriation, "and it finally scratched through by a very narrow shave." Oliver relented only when he was assured that the appropriation would be the last one for this research.[60]

Metzger pressed on, despite mounting opposition. "The truth is that this smelter problem now appears to be a permanent one," he wrote in early 1935. "Developments in the last few years indicate that the smelter cannot operate profitably without causing injury in the United States." Efforts to get the Canadians to impose an operating regime on the smelter had so far been unavailing. The fixing of a specified maximum level of pollution across the border, involving as it did the famous yardstick question, was vigorously opposed by the smelter. And as long as the smelter was emitting fumes, and if no budget were granted for the investigation, "we shall be left without any evidence or means of meeting the contentions of the Canadians."[61]

In 1935, Senator Kenneth McKellar of Tennessee informed the State Department that "there is great opposition to the appropriation for the coming year [FY 1936] and the Committee has instructed me to write to you that it does not intend to make any further appropriation for this purpose." Why not just compensate the farmers outright? he asked in the hearing. Some $150,000 had already been spent with no end in sight. Yet, as Metzger pointed out, the other side had spent a great deal more than that and would continue to do so. The $19,000 being requested would cover the salaries of a senior pathologist (Hedgcock), a chemist (Griffin), and two junior chemists, but there were no funds for a plant pathologist (Goldsworthy). "The fact is, they have curtailed the investigation to the very minimum," Metzger testified, and "as it is now, we are not able to do any experiments with fumigation. When we had the $40,000 a year, they conducted experiments with vegetation to see what the effect of certain concentrations was on vegetation, but we had not been able to do that since we have had this small appropriation." Washington senators Clarence Dill and Lewis Schwellenback supported the appropriation, but it was cut to $15,000 in FY 1936.[62]

Year after year, these difficult hearings had the effect of a thousand cuts on the USDA's investigation, and industry cannot have been unhappy with the result. In 1936, Arizona's Carl Hayden and Nevada's Wright Pittman were advocating the USBM's position on smoke controls at hearings of the Senate Appropriations Subcommittee; "That is the only remedy," Pittman said. And if the State Department were to secure a big judgment against Trail, Hayden warned, U.S. smelters at Douglas and El Paso would be liable for the same thing by Mexico.[63] Political support for the USDA and its controversial research agenda was fast fading away.

With the conclusion of the first set of hearings before the Arbitral Tribunal in 1938, no further appropriations for fumes research were requested. The second award, as has been discussed, was minuscule, $78,000, when more than $1 million had been asked for. And while the State Department was willing to reopen the case, the Smelter Fumes Committee felt that this action would "not be desirable from the standpoint of the Department of Agriculture regardless of what we might think of the decision as there is a question whether we could receive any credit for such action." Knight looked back on the ten years of this research as an example of effective interbureau cooperation within the department. "Unfortunately," he concluded, "during much of the period we were handicapped for funds and there was a lack of sympathy expressed in certain quarters for the group making and asking for redress."[64]

The Smelter Fumes Committee evidently met for the last time on May 8, 1938. Having directed the case from its inception, Skinner was appointed master in chancery to distribute the combined IJC and Arbitral Tribunal awards to the farmers. George Hedgcock, past retirement age, agreed to stay on without pay to pull the scientific papers together for publication. But the active research effort was closed down, and the committee disbanded. The papers were never published.

In Ottawa, meanwhile, the Associate Committee on Trail Smelter Smoke held its eleventh and last meeting on November 23, 1937. A letter of appreciation from James J. Warren was received, along with an offer to pay honoraria and to cover the staff time needed for Morris Katz to prepare their papers for publication. Copies of Warren's letter were sent to the Departments of Agriculture and Mines as well, indicating the high degree of interdepartmental cooperation that existed—so unlike the bureaucratic elbowing between USDA officials and the USBM.[65]

Flushed with the excitement of new discoveries, Skinner and the U.S. team had been eager from the start to publish their results, and discussions began as early as November 1929 on the scope and purpose of such a work. Initially, the State Department lawyers wanted a report in nontechnical language for use by the IJC commissioners. Two years later, with the IJC hearings over, the Smelter Fumes Committee recommended a somewhat different volume based on a long, general introduction, followed by the scientific papers. Skinner had been opposed to premature publication before all the data were in; Kellerman was eager to publish and to claim primary credit for his Bureau of Plant Industry. The agreed upon solution was to publish the results of four years of research up to December 1930. To delay further was risky, because the Canadian team was then very active at Summerland, doing "an extensive series of fumigation experiments in which they are overdoing themselves in the way of the most expensive, elaborate and spectacular equipment which money may procure. There is no telling what they may publish," Griffin worried, "thus 'scooping' us . . . and usurping the place we should have in the literature at the conclusion of this investigation."[66] In the end, they were scooped by the Canadians.

Following the IJC ruling in 1931, awarding $350,000 in damages to the farmers but with no decision made whether to accept it, the U.S. scientific team felt vindicated and anticipated publishing its results in a lead series of three articles on the effects of SO_2 deposition in small amounts on vegetation and its significance in smelter fumes. Griffin and Skinner did publish the first paper in 1932 in the *Journal of Industrial and Engineering Chemistry* under the title "Small Amounts of Sulfur Dioxide in the Atmosphere: An Improved Method for the Determination of Sulfur Dioxide when Present in Very Low Concentration in the Air." In their abstract, the authors pictured themselves as heirs to the German scientists who had established the new science of air pollution but lacked the precise measuring instruments of the sort now available to support their work as they described it. Nor did the Selby investigators have access to the new monitoring equipment. This claim was most clearly and provocatively an opening salvo, and M. D. Thomas in Salt Lake, inventor of the portable recorder improved upon by Griffin, responded immediately with a sharp letter to the editor, attacking them for not giving him credit and questioning the significance of their "minor modifications" to his recorder. This return salvo was most obviously from ASARCO, implicitly casting doubts on their line of research. Griffin had been aware for some time that Thomas and he disagreed "in respect to many important phases of the sulfur dioxide–air contamination profile, but I had not expected that his criticisms of our work would take the form assumed in his letter [to the editor] dated August 13, 1932."[67]

In June 1933 Knight submitted with urgency to the Smelter Fumes Committee a list of some thirteen papers for publication, probably this time as department bulletins because the Bureau of the Budget was recommending zero funding for FY 1934. (The list of topics was paralleled in the Canadian volume, published in 1939.) The first paper in the proposed series, on the "Accumulation of sulphur in foliage exposed to smelter fumes," was submitted by Fisher and Griffin to the State Department for review in February 1934. However, Metzger asked them to withhold it, and the other papers too, from publication until after the dispute was arbitrated, as premature publication might tip the hand of the U.S. negotiating position. Furthermore, there were "a few items in the manuscript which would be unfavorable to our side of the case and which we are under no obligation to divulge," Metzger instructed Knight. Having worked on the case since 1927, Fisher was greatly disappointed. "A great mass of most interesting and valuable data has been accumulated and I believe that for the first time adequate evidence from the agricultural as opposed to the smelting standpoint has been obtained. I feel very strongly that it should be published." But Knight and his colleagues could do nothing until the State Department decided to release these papers, and the series petered out. Clearly, Metzger regarded the papers first and foremost as evidence to bolster the legal case for controls on the smelter and a higher damage award, whereas the scientists were eager to contribute to the literature as well as aid the farmers.[68]

Given the pared-down appropriations and with the State Department calling the shots, the most that the Smelter Fumes Committee could hope for was to publish a few articles but not a book or a series. Entries in Knight's diary dealing with the publication issue ceased after January 1934. Furthermore, the climate in official Washington had turned against them. Reportedly, some hostile letters were being received by the professional journals, indicating that their work would not be treated kindly by scientists friendly to the smelters, which meant most of the leading experts on air pollution. The articles that were published in the mid to late 1930s—by Thomas and Hill, Swain and Johnson, and others—bolstered the other side of this scientific dispute.

Well after the Arbitral Tribunal's ruling in 1938, Griffin and Skinner were still discussing a series on air pollution, containing Griffin's more recent work on the contamination of rain and snow, but nothing came of it. War work intervened, and Griffin had the unfortunate work habit of always being behind in preparing his data for publication. Aside from their jointly authored paper in 1932, Skinner himself never published again in this field. As late as 1942, the elderly George Hedgcock was preparing an answer to the 1939 Canadian volume. However, the most that ever came of this effort was a Forest Service technical bulletin that appeared posthumously in 1955. In fact, excepting Hedgcock, none of the USDA men went on to make a name for himself in "this most difficult and illusive problem."[69]

In contrast to the decelerating American research and publication effort, the Canadians enjoyed a surge of achievement and recognition, culminating with the 1939 publication of *The Effect of Sulphur Dioxide on Vegetation*. The NRC had intended to publish in the *Canadian Journal of Research*, but C.M.&S., well pleased with the way the case was going, offered to pick up the costs of editing and printing a hardcover volume. Katz, Ledingham, and McCallum did most of the writing and editing and were instructed to proceed immediately. With this final instruction, the Associate Committee on Trail Smelter Smoke folded its tents in November 1937.[70]

In the absence of publication, what did the USDA scientists uncover, and were they close to making a breakthrough in the early science of acid deposition? Certainly the smelting companies thought they might be and by 1935 were mounting an all-out effort to discredit their work. In particular, USDA researchers thought they had gone beyond the cabinet fumigation work of O'Gara and Hill at the ASARCO experimental station in Salt Lake to discover strong evidence of continuous, invisible injury to plants under long-term exposure to low concentrations of SO_2.

Arguing in 1934 before the House Appropriations Committee that the field research and the recording stations must continue, Skinner testified:

> The unfavorable and destructive effect of fumes emitted from the smelter at Trail, BC, upon the vegetable and animal life in the upper Columbia River Basin . . . has been studied and the character and extent of the destruction determined. Data have been secured which permits a definite prognostication of the complete destruction of the area affected unless the nuisance is materially abated or stopped.
>
> The damage and destruction are continuous and accumulative, and research has shown that the area affected by the fumes is continually expanding, although this conclusion is challenged by the smelter company.[71]

Skinner was putting the best face on a complex case, not all of it favorable to the United States, even in 1934. For example, the acid recovery plants combined with Consolidated's development of a weather-coordinated plume control regime in 1934 would make it more and more difficult to sustain the last part of his statement. Until then, the NRC team had been working in the shadow of an awkward fact: as Katz related, "reports on frequency, duration and intensity of fumigations for 1932 and 1933 do not show the improvements with respect to atmospheric pollution which were apparently anticipated by the Commission [the IJC]."[72] However, by 1934 these acute burns were being brought under control. But what of the "chronic, cumulative and permanent injury" that the United States claimed was still occurring in Stevens County?

In 1937 Bill Skinner briefed Swagar Sherley, the new U.S. agent replacing Metzger who would soon be presenting the U.S. evidence in hearings before the Arbitral Tribunal:

> In this case, the only proven and, in my opinion, the only provable cause of damage is sulfur dioxide. Furthermore, the experimental

work in this case has shown, I think, conclusively (although I am aware that it is challenged and will be challenged in the future) that plants subjected to intermittent and continuous fumigation with sulfur dioxide acquire a supersensitiveness to the toxic effect of this compound. The experimental work does show that plants subjected to intermittent fumigation with sulfur dioxide accumulate within the cells sulfur compounds the exact nature of which is undeterminable and unknown. The best chemists can do is to show that the total sulfur has been increased in the plant or in the plant ash. This has been conclusively demonstrated in the experimental work with more than a dozen species of plants.

He then reached the core of the position Metzger had understood and would have marshaled in the hearings had he lived: "The crux of this case is not demonstrating the exact measurable damage of any particular fumigation at any particular time, under any particular conditions of moisture or temperature, but that continued intermittent fumigation does result in chronic damage, which in some cases results in destruction."[73] Work at Wenatchee showed that the feeding value of barley, hay, and alfalfa was reduced—farmers in the smoke zone were planting less and less barley, the grain crop most susceptible to injury.[74] Consequently, the key question was not the amount of damage as determined by visible markings on the crops—the company's position—but the effects of cumulative and invisible injury that were causing economic losses.

To be sure, the U.S. experimental data to back this claim up were already several years old when the hearings began. Katz had fresh data from the 1935 and 1936 seasons at Summerland that were damaging to the U.S. case. The effects of cumulative injury, as claimed by Hedgcock, could in fact occur as a result of the increase in sulphur content of the foliage of evergreen trees over a period of time. But trees at Summerland were fumigated only once, so that the Canadian experiments did not demonstrate this form of injury. With respect to invisible injury, experiments on alfalfa showed that "there is such a form of injury which can be briefly described as a reduction of photosynthesis caused by the presence of sulphur dioxide." However, "this is a temporary condition, the plant activity returning gradually to normal with the removal of the gas. This adverse influence on photosynthesis has not been found to occur with concentrations under 0.40 to 0.50 p.p.m." No evidence was found that injury was progressive: even in the winter, under conditions of high humidity, once the gas was removed the rate of photosynthesis "rapidly returned to normal." In addition,

during the summer months, with bright sunlight, high temperature, and low humidity, the initial symptoms of injury were much more readily detected and there was no basis for the suggestion of "invisible injury" such as was described by the Americans. Also, the development of symptoms subsequent to injury was much more rapid than during the winter months, showing that this was caused by external conditions rather than by sulphur dioxide.[75]

The questions raised by this newer Canadian data weakened, if not disproved, the U.S. case. But the immediate effect was on the three judges at the tribunal who, in their annoyance at the inability of the two teams to coordinate their experiments and cooperate, saw the scientific side of the case evolving into a typical, polarized, smelter dispute.

As well, Sherley had only a shaky grasp of the U.S. argument—in fact, he came on the case proclaiming that the USDA had no case—and bungled the disastrous July hearings in Spokane while earning the contempt of Knight and his colleagues.[76] Bucked up by Skinner at the September hearings in Ottawa, Sherley and the U.S. team did better at that time, but the damage to their credibility had already been done.

Early on, Read had identified a fundamental weakness from the legal perspective: "The U.S. claim to damage and proof and measurement seems to lie entirely on [the economic survey done by Sanders]." The Canadians discredited Sanders's work by showing that his damage surveys were based in part on oral testimony by interested parties and did not correlate with area statistics gathered by the U.S. Census Bureau. They poked holes in his methodology, and Tilley pressed home the attack. Furthermore, Tilly said, after purporting to show the existence of injury, the scientific surveys just stopped.

It is true that they talk about severe injury and percentage of injury, but on the other hand they do not present any evidence of a quantitative character, nothing upon which a tribunal could make a finding that so much alfalfa or timber could have been injured to such and such a degree, involving an award of a specific number of dollars. Consequently, it seems to me that in so far as the money claim is concerned, the U.S. case depends entirely, and stands or falls with the Sanders Report.[77]

The U.S. assertions about chronic, invisible, and economic damage were either not proved or in many instances were disproved by their own countervailing scientific evidence, the Canadians said. Checked on the scientific side, the American claim for damage was checkmated on the economic side.

The smelter was less successful heading off a finding of the tribunal that additional remedial measures were necessary to prevent occasional burns; a more stringent regime was imposed but without a rigid yard-stick. Nor did management escape having to pay the additional costs of these stricter emission controls, a direct result of the prolonged U.S. effort to force the smelter to cease polluting across the border. An innovative supplemental control regime (SCS) was imposed on an experimental basis between 1938 and 1941 by Swain and Dean, the technical consultants to the tribunal.

Based on research conducted by Swain and George Hewson of the Canadian Weather Bureau, who advanced the meteorological understanding of plume behavior and control, the new regime in large part solved the acute injury problems caused by sudden downdrafts and the formation of local hot spots. In 1941, this regime was made permanent by the tribunal. However, because it was tailored to the specific problems of Trail emissions, no generalized rule was intended—to the smelters' relief. As the consultants wrote in their report, "a complete solution to the problem of damage to vegetation in the United states can be reached without an arbitrary limitation on the total amount of sulphur dioxide emitted by the smelter."[78]

The USDA fired a parting shot of its own. Following the tribunal's disallowal of most of the U.S. claims, its suggestion that S. W. Griffin, the chief chemist, do a joint crop survey was turned down flat by Griffin, who did not want to work with smelter scientists or the two technical consultants (Swain and Dean), whom he considered pro-smelter.[79] Instead, Griffin was instructed to stay in Northport, where he could check on the regime's effectiveness by observing crops and taking fumigation readings for 1938, 1939, and 1940. Griffin also took strong exception to a conclusion in the Dean-Swain report which affirmed that, based on field reports by Swain's former graduate student, now an employee for ASARCO, there had been no plant damage in these years. He so testified before the tribunal.[80] And when the Bureau of Mines sought to publish the Dean-Swain report in its regular bulletin series, thereby replacing Selby with a new industry standard, the State Department intervened to change the document.

The bulk of the Dean-Swain report dealing with purely meteorological observations was indeed "a meritorious contribution," as Secretary of State Cordell Hull wrote Secretary of Interior Harold Ickes in 1941. However, at the USDA's request,

> that part of the report . . . which relates to so-called investigations of plant life south of the international boundary, which, it is understood, is largely the product of Dr. Swain, and in which it is contended that no damage of any consequence was caused during the years 1939–1940 inclusive . . . [reaches conclusions] at variance with the official position of this Government, [and] should not be printed

as a publication of this Government unless accompanied by a reproduction of the documents and arguments presented by the U.S. Government in support of its official position.

Interior complied; Swain duly "removed all controversial matter"; and
the Bureau of Mines proceeded with publication, proclaiming, "This report will replace the Selby Commission report as the new standard reference work on Smelter Smoke."[81] Yes, the atmospheric studies by Swain
and the Canadian Weather Bureau led to new techniques for emission
control. But the USBM Bulletin did not contribute anything new to the
debate on phytotoxicity.

Noting that "certain liberties" had been taken in the published version, especially "the complete deletion of all field surveys conducted by the
consultants to determine damage to vegetation, and that the conclusions
from these surveys were extremely favorable to us," the NRC pondered
what to do about it. Read advised that it would be undesirable to lodge a
protest with the Bureau of Mines. "There may be trouble ahead of us after
the war and I think that it would be unfortunate to antagonize Dr. Dean
who has probably acted in good faith in the matter." Instead, he suggested
that the matter could be dealt with in a review, published in a Canadian
scientific journal, that "would avoid offence to Dr. Dean and Dr. Swain and,
at the same time, the points raised [by the NRC] could be corrected."[82]

Swain, after all, was a valuable, credible, and self-assured ally. Writing to Read after a site visit in 1943, Swain concluded:

The regime laid down by the Tribunal, and ably administered by the
Smoke Control Office . . . is operating with marked success. No damage whatsoever by emanations from the smelter is being done across
the international boundary. Careful field studies are fully supported
by the data on sulphur dioxide concentrations obtained from the
recorders. Such a clear record in these times, when wartime production is being pushed necessarily to the highest possible level, within
the limits imposed by the regime, is most encouraging. The regime
has met with its supreme test in these years of war. With the return
to a wider degree of freedom in peace-time smelting operations, and
with further advances in smelter practice and by-product recovery
which are bound to come in the years ahead, sulphur dioxide from
the plant of the Consolidated Mining and Smelting Company should
never again become an agent of injury to plant life in the Columbia
River basin below the international boundary.[83]

In light of the fact that Consolidated was trying to understand why certain unexpected fumigations were occurring, the reason why Professor

Swain was up there once again (see Chapter 3), these were helpful words indeed. Even "Mr. Griffin's weeds," as Swain referred to the evidence used by Griffin (with a reference to his M.A. degree) to attack the Dean-Swain report, were free of any signs of injury.

The last word on plant injury was expected to be the NRC's *Effect of Sulfur Dioxide on Vegetation*, which was printed in early 1939. Capably organized by Katz, McCallum, Ledingham, and Lathe, the 447-page volume features the scientific evidence that led them to conclude that the invisible injury thesis could not be sustained, whether through the effects of sulfur accumulation in leaves or leaf function or declining yield. Fumigations below 0.30 ppm caused no injury over prolonged periods, and plants exposed to higher dosages up to the threshold of acute injury could recover, although damage could at least occasionally result.[84] By contrast, the unpublished USDA reports languish still in the archives.

In Canada, the principals received honoraria from the company as well as letters of citation from the prime minister in 1937 and later from General McNaughton. Morris Katz became a distinguished air pollution expert, in 1969 receiving the R. S. Jane Memorial Lecture Award from the Chemical Institute of Canada, of which K. J. McCallum was then president. Frank Lathe, who was also a past president of the Chemical Institute, received the OBE shortly after leaving the NRC and was consultant to several mining and industrial concerns before his death in 1960. Unheralded, the U.S. government scientists went on to other work.

Morris Katz reiterated these results at a postwar symposium on air pollution, organized by Swain at a meeting of the American Chemical Society in San Francisco. Controlled experiments at Summerland showed that various plants could recover from the effects of low concentrations of SO_2 in the absence of visible injury. Only with leaf destruction was there damage. He wrote:

> In none of these investigations did the treatment decrease the yield, photosynthesis, or respiration, or interfere with the march of the stomata, or deteriorate the chemical composition, protoplasm, or other plant structure. A reduction in yield or rate of growth was correlated with a definite amount of leaf destruction or loss of foliage in fumigations of sufficient duration at concentrations above 0.25 p.p.m.

He then concluded: "It is hoped that the 'invisible injury' theory has now been disposed of, once and for all time, and will not be resurrected again in problems involving sulfur dioxide damage to plant life."[85] With this parting shot the controversy died down. Interest had already declined with the Trail settlement, emphasizing as it did control of pollution through a combination of technical and dispersion methods.

In hindsight, USDA scientists produced innovative work that was suggestive, like the German research before them, but not yet conclusive. Still, in reviewing their work and their hopes, one sees the limitations. For one thing, the U.S. effort, unlike the Canadians', was dissipated. For another, the standards of legal proof took pride of place over science. In short, the definition of damage was narrowed to exclude their findings on cumulative, invisible injury. These findings were controversial and were challenged vigorously by the Canadian team. For still another, the balance of forces arrayed against the American team overwhelmingly favored industry. In the calculus of public utility, the Trail smelter, the largest such complex in the British Empire, counted more—for its jobs, for its contribution to the Canadian economy, for its essential war production—than the claims of a few hundred poor farmers engaged in marginal agriculture. Hence the prevailing intellectual framework of this dispute favored a technical solution. With this factor in mind, in the next chapter I will discuss the role of industry and the international implications of the Trail case.

CHAPTER THREE

Paths Taken and Not Taken

"Process industries should begin to worry again about 'smoke farming,'" a trade journal editorialized in 1937. "This worry is rightly inspired by recent ultra-zealous efforts on the part of certain agricultural specialists in Washington." Even though the Smelter Fumes Committee and the USDA field staff were now largely inactive because of budget cuts and lack of support in official Washington, the issues they had raised about the effects of low-level SO$_2$ fumigations on crops and forests were still pending. "Unless some interruption to the present course of events can be arranged," the journal continued, "the Trail settlements may establish a dangerous precedent for other industrial enterprises in the States, particularly the smelters."[1]

What the smelters most wanted to finesse was "setting a rule as to the frequency, duration and amount of concentration permissible without causing damage or injury."[2] This was the famous "yardstick question" that researchers at ASARCO, the U.S. Department of Agriculture, and Canada's National Research Council were all attempting to pin down, with divergent results. The assessment of liability depended on which school prevailed, the proponents of invisible injury, or subtle and accumulative damage, whose thesis was under heavy attack from industry, or those maintaining that proof of damage should be limited to the visible signs of foliar injury. However, as has been discussed, the Americans failed to sustain their interpretation of the scientific evidence at the tribunal's hearings in 1937, which reinforced the tribunal's preference for a technical solution to the fumes problem. Frustrated by the inability of scientists to cooperate, let alone agree on the results of their parallel investigations, the tribunal sidestepped making a rule on the scientific dispute.

Although its importance as a landmark case in the evolution of international law on transborder pollution has long been heralded in the literature, questions about the actual significance of the prolonged fourteen-year-long Trail smelter dispute have been raised by legal scholars.[3] To be sure, Trail did extend the "polluter pays" principle to cases of transboundary pollution, but this result was a by-product; the main issues of the case have been otherwise ignored or obscured by the conventional wisdom about Trail's significance. It is the historian's task to establish a context while sorting out the evidence in order to find the underlying meanings embedded in the abundant records of this complex case. Why was it that, for all of the money spent on scientific investigations and the hearings themselves, the actual results in terms of both the precedent for damage recovery and the legal standard for pollution abatement were both so narrow? One reason why a case that promised so much in fact delivered so little was because the issue of actual damage was so narrowly defined.

The collapse of the American scientific case left the field to Read, the very able Canadian agent, to Tilley, the seasoned lawyer for the CPR, and to Dean and Swain, leaders in their respective fields who devised the SCS regime. "The Tribunal's problem all along," Read observed as the hearings wound down in 1941, "is dealing with the problem of assessing damages," which could only be done by actual observation in the field, to verify decreased yield. But if the damage were to be awarded on a standard based on ppm data from the automatic recorders, "this would be a heavy charge, an automatic charge, on the smelters, into the hundreds of thousands of dollars." The United States had wanted a regime "by which a prefixed sum would be due whenever the concentrations recorded would exceed a certain intensity for a certain period of time or a certain greater intensity for any twenty-minute period."[4] Since 1934 the Americans had been pressing for such a standard based on the claim that invisible injury was being caused by low concentrations of smoke from Trail in the 0.10 to 0.20 ppm range. Tilley then drove home the point: it was never intended for the tribunal "to give indemnity for anything more than the direct damage caused by fumes"[5]—not for falling land values, municipal taxes forgone, the inability to secure credit, and certainly not for so-called invisible injury.

The other side of this argument was crafted by W. W. Skinner, leader of the U.S. scientific team, who pressed the case for cumulative and progressive damage caused by the effects of repeated fumigations of sulfur dioxide on plants. In a letter briefing Metzger's successor—Swagar Sherley, a lawyer and retired congressman doing a job he neither relished nor particularly believed in—Skinner explained the core of the argument that Metzger understood and would have pressed on the tribunal: "The crux of this case is not demonstrating the exact measurable damage of any particular fumigation at any particular time, under any particular conditions of moisture or temperature, but that continued intermittent fumigation does result in chronic damage, which in some cases results in destruction."[6] But then Metzger died, leaving no notes or instructions on how he would have argued the case. Swayed instead by Dean's briefing on the technical aspects of emissions control, and with little confidence in the results of his own government's scientists, results he could not understand, Sherley then bungled the hearings.

Nonetheless, the famous yardstick question was addressed, if indirectly and not in a way the American scientists had intended or preferred. The parties did agree that direct damage had been declining since 1936, at first when the smelter installed its own control regime, and then in the years 1938–1940, when the tribunal controlled emissions under the experimental regime installed by Dean and Swain. Noticing a dramatic drop in the readings of SO_2 concentrations of ppm/hour on the Northport recorder, Stewart Griffin wondered if "the smelter management itself is finally and inadvertently providing a standard or reference value (in respect to future sulfur dioxide emissions) which would be satisfactory to us and which demonstrably they can meet and which, in fairness and reasonableness they may be expected to meet and yet operate their works." In the past. U.S. suggestions for establishing what Griffin called "limiting air-pollution values [in the range of 0.1 to 0.2 ppm] such as might be tolerated in this vicinity" had been rejected by the company as entirely too low for practical smelter operation. But now that the company itself was achieving "a practical standard," why shouldn't this standard become a requirement, wrote Griffin to Skinner on the eve of the hearings in Spokane—the July 7–29 hearings that turned out so disastrously for the USDA, and where the settlers lost all hope of a large damage award.[7]

The tribunal mandated Trail to cause no injury, and to this end a daily regime was established to curtail burns that could cause acute, visible injury. For control purposes, a recorder was maintained at Columbia Gardens, giving a continuous record at the smelter by wire. After fine-tuning, the regime was made more restrictive, especially during the growing season. Subject to meteorological conditions, the standard sulfur base load was established at four tons per hour from 3 A.M. to 9 A.M., and seven tons per hour during the rest of the day and night, one ton of sulfur being the equivalent of two tons of sulfur dioxide.[8] At 100 tons per day (maximum, from April into July, the most sensitive period for certain plants and trees), this sulfur balance standard was a substantial reduction from the 300–350 tons of sulfur emitted daily in 1930.

If the Columbia Gardens recorder indicates 0.3 part per million or more of sulphur dioxide for two consecutive twenty-minute periods during the growing season, and the wind direction is not favourable, emission shall be reduced by four tons of sulphur per hour or shut down completely when the turbulence is bad, until the recorder shows 0.2 part per million or less of sulphur dioxide for three consecutive twenty-minute periods.

In the nongrowing season, concentrations of 0.5 ppm would be permitted under the same regime, also depending on wind velocity and direction.[9]

Under this regime, the Canadians noted, there had occurred on the Northport recorder a total of just 5.7 hours of SO_2 readings above 0.25 parts per million during the entire growing season of 1941. "This compares with 2.0 hours in 1940, 7.7 hours in 1939, 12.1 hours in 1938, 5.3 hours in 1937, and 50.8 hours in 1936." To be sure, between readings at Columbia Gardens just north of the border and Northport, a few miles south of the line, the incidence could be expected to decline somewhat. Nonetheless, "there were no readings above 0.50 parts per million at the Northport recorder in the growing season of 1941."[10] However, in the winter of 1941 burn markings were noticed. Then, in the winter of 1942–1943 readings were some 20–30 percent higher at a recorder below the border, and burn markings were again being noticed a good thirty miles below the border. Read asked to be notified immediately if a serious burn occurred, but the problem seemed to be lighter burn traces, for which there was as yet no satisfactory explanation, and farmers were noticing these traces.

In fact, making "the correct interpretation of light sulphur dioxide markings on vegetation is extremely difficult," F. Lathe, the head scientist at the NRC, remarked, even for their man with the most field experience in the area. Moreover, "the scientists of the USDA, who have been working on the case for ten years or more, have insisted that, even without visible markings, substantial damage is sometimes caused." There was concern that the whole question might be reopened and production curtailed—if the Americans were to show that "fumigations had far exceeded anything permitted [by] the Tribunal, and to now urge that the liberation of sulphur dioxide at Trail be radically reduced."[11] Such fears proved unfounded. The cause for these anomalous recorder readings was the effects of unusual atmospheric conditions rather than increases in emissions owing to war production. Following a site visit in 1943, Swain reassured Trail management by observing that the regime had "proved itself by the results achieved, and should not be changed in any particular."[12]

Would more stringent values have been established if the invisible injury thesis had been upheld? Only marginally, it would seem, which was a victory for the Americans, although by the back door. By 1936 Hedgcock, the forest expert, was noticing improvements in his three zones. Later, in what turned out to be the only report on the American scientific work to appear in the published literature, he acknowledged that the damaged landscapes around Northport had substantially recovered under the new, more stringent regime.[13]

Designed to prevent direct, visible damage to plants, the SCS was tailored to meteorological currents in the river valley. It was entirely consistent with Hill's and Thomas's findings at the ASARCO station in Salt Lake that plants could tolerate fairly high thresholds of exposure to SO_2

without showing signs of visible injury. But under the new regime, Swain claimed, no damage had been done in the Northport area, and the problem was resolved. Griffin challenged this claim (which resulted in the final version of the USBM report being published without Swain's observations on the absence of damage, as has already been discussed). The beauty of SCS was that it was based on a quite generous yardstick that, in conjunction with production cutbacks when warranted by meteorological conditions, did reduce pollution substantially. But no new generalized standard to replace Selby came out of the Trail case, and the lawyerly approach to damage had the effect of keeping the issue on traditional and safely demarcated grounds. Thus the tribunal left open the complex question of plant injury, but the urgency had gone out of it. This nonruling had the effect of stifling basic research on lines and in directions that the industry did not like.

It is now known that SO_2 at low concentrations over weeks or months can in fact injure sensitive vegetation, despite a number of sulfur dioxide studies done earlier to disprove the invisible injury thesis. In Walter Heck's concise definition (1989), the effects of ozone and SO_2 on crops "is either visible or subtle [i.e., invisible]. Visible effects are morphological, pigmented, chlorotic, or necrotic foliar patterns resulting from major physiological disturbances in plant cells. Subtle effects are measurable growth or physiological changes without visible injury that may affect yield or reproductive or genetic crop systems."[14]

A characterization made decades earlier by the Germans, that visible injury is either acute or chronic, has been long accepted. But the invisible injury thesis has been accepted only recently. Near the end of his career Thomas himself was finding some evidence for Stoklasa's postulate—invisible injury defined as interference with the growth or functioning of the plant without attendant lesions on the leaves or elsewhere—and went so far as to say (1961) that "possibly a fumigation treatment could be devised that would produce a significant amount of invisible injury. But, he concluded, "it would probably be of academic interest only," since the problem was so unimportant in percentage terms relative to damage indicated by visible markings. However, by the 1970s research on damage around power plants and the onset of acid rain studies showed otherwise. "While acute sulfur dioxide toxicity has received most of the attention in the past in smelter areas," the University of Toronto's Thomas Hutchinson, a botanist, noted, "a lot of attention is now focused on chronic or low-level sulfur dioxide effects." Two East German scientists reached similar conclusions in their 1988 book, *Air Pollution and Its Influence on Vegetation*. In short, the earlier work of Thomas, Hill, and Katz was not the last, or latest, word. And in a recent text (1997), Sagar Krupa defines chronic response as "yield effects without symptoms," that is, what used to be called invisible injury to plants is still, he affirms, a concern today.[15]

For assessing yields, the problem is still what it always has been: acute foliar injury is the only good field indicator of cause-effect relationships, according to Krupa. Since chronic foliar injury may be mimicked by other biotic and abiotic environmental stresses or may resemble norman senescence, it is not a good field diagnostic tool. From a practical standpoint, Krupa concludes, visible foliar injury is the only conclusive way to identity O_3 or SO_2 injury in the field. For this reason, well into the 1980s, that is to say, two generations after the Trail smelter case, visible injury in its acute form was still the only generally accepted legal standard of damage to crops from point-source emitters like smelters. From the advantage of hindsight, the USDA was assigned an almost impossible task.

But the complexities of this research alone cannot explain why the early work was not followed up and a generational gap occurred, so that even the memory was lost. It is probably no coincidence that a similar outcome occurred in Germany, where the science of smelter pollution began. Engelbert Schramm, who has studied the environmental history of German air pollution, concluded that the work of Stoklassa and Wislicenus was not carried on with the same intensity by the postwar generation. In part, it was because the depressed economic conditions of the 1920s were not conducive to supporting research in independent laboratories. But in part also it was because industry developed a legal standard for damage awards that depressed the demand for further basic research. Industry employed scientists for narrower, instrumental ends. In other words, the emphasis in the courts on establishing cause and effect discouraged work in basic science. Leads on the synergistic effect of various combinations of pollutants, for example, SO_2 and lead dust, were not followed up. Investigating the interplay of different absorption mechanisms in plants did not fit the legal requirements for precise, direct proof of damage—for which industry wanted to defend itself against court challenges from the farmers and to argue against pollution controls. "Ultimately, these types of debates—this [original] way of framing the question—did not go forward because the majority of the scientists were not prepared to take indications of complexity seriously."[16]

In his history of smoke pollution, Gerd Spelsberg also affirms that the generation after Wislicenus was influenced by the juridical and economic points of view, which led to a restricted and narrow definition of damage. Experts moved away from basic science to consulting. In fact, the notions of Stoklasa and Wislicenus on invisible jury were a lot more complex and imaginative than the new juridical focus on plant damage. There was a generational shift away from the pioneers, who worried about the general effects of pollution, to an increase in tolerance for air pollution and a narrowing of interest in the subject, eventually to the loss even of the memory of the earlier research until concerns for the environment revived in the 1950s and 1960s.[17]

In this same period dispersion was favored in Germany as well. What happened in the heavily industrialized Ruhr was a race to build higher stacks in order to diffuse emissions over a wide area. In consequence, Spelsberg writes, a blind belief in the capacity of nature to purify and cure itself developed.[18] In the Nazi period alone, some twenty-eight papers dealing with dispersion and the measurement of SO_2, CO, and dust were published.[19] But war production took pride of place, and there were days when the smoke was so thick that Allied bombers could not find their targets in the Ruhr. In fact, only in the 1950s did German approaches to pollution begin to shift from an emphasis on end of the pipe strategies based on diffusion technology to looking at the problem holistically once again.

Similarly, in the United States there was little interest in researching the effects of SO_2 on crops until the 1970s, when new interest was stimulated by long-term assessments, especially around power plants. These results "convinced the scientific community that earlier perceptions of SO_2 thresholds were flawed and would have to be reexamined."[20] This reassessment would include the NRC papers, published in 1939 as *Effects of Sulphur Dioxide on Vegetation*.

Thus by the 1970s, science and government were again concerned with the overall problem of SO_2's effects on crops in the vicinity of sources. Air quality regulations and the application of control technologies then mitigated this problem in the 1980s. Still of concern today are the joint effects of SO_2 and other air pollutants, such as ozone and nitrous oxide, and the long-range transport of SO_2 and NO_2, which, breaking down into sulfate and nitrate compounds, are the building blocks of acid rain.

When all is said and done, the German and the North American experiences are too similar to be coincidence. Just as ideas and technology are highly fungible across international borders, so too do the techniques of dealing with the consequences of technology quickly find their way to those who need such knowledge, in this case to the smelters fighting tighter emissions standards. In North America, the mine and smelting industry has a tradition of sharing technology. The conclusion must be that the flow of knowledge was a transatlantic phenomenon as well.

In reviewing documents including bibliographies generated by the U.S. scientific program, I noticed that the USDA researchers were not availing themselves of a more recent German literature than Soraurer or at best Stoklasa. Nor, by the 1930s, did Swain and his graduate students refer to current German research in their work.[21] Nor did Canadian scientists at the NRC reference current German work. The reason is that basic research dried up in Germany *just before* the Americans needed more confirming evidence of what they were finding at the USDA research station in Wanatchee. It seems that USDA field scientists were already out of step with a basic trend in the linked worlds of industry and air pollution science. They were still looking for invisible injury, whereas most people in Germany and North America by then had taken a quite different pathway to pollution abatement, that of technical solutions and diffusion, under which industry could still operate profitably.

At Trail, the control regime functioned smoothly despite the demands of war production. Low-level fumigations into U.S. territory continued, accompanied by a visible fine mist or smoke which was sometimes quite heavy. Although not harmful to vegetation, a company scientist observed, "It gives the layman a chance to blame smoke for many of his troubles," and efforts to correct the conditions causing visibility problems were continued.[22] But the combination of acid plants, SCS, and diffusion worked well by the standards of the times. As F. E. Lathe, the NRC's head chemist, wrote to Reginald Dean, "The SO_2 problem is still an active one with respect to conditions at other Canadian smelters. At Trail, as you are probably aware, all available SO_2 is being used in the manufacture of sulfuric acid, and, as a result, the amount being voided to the atmosphere is much lower than for many years past."

In reply to an Ontario judge inquiring about the literature on smoke controls in the context of SO_2 being vented from the nickel works at Sudbury, Canada's worst polluter, Dean summed up the current consensus on how to deal with smelter pollution:

Due to the huge economic damage which could be done by shutting down such a smelter or by saddling it with requirements to mitigate its damages by installation of expensive changes or additives, any move made should be carefully considered. The metallurgical industry itself wants to become a better industry and to recover all waste products that can be recovered economically. Doubtless all authorities want to see any changes recommended so designed as to increase rather than decrease the economic output of such an enterprise. The Trail Smelter in British Columbia has gone far in this direction and can probably serve as the present living model on just how to mitigate smelter smoke nuisances.[23]

The first lesson to be learned from a close reading of the Trail case is that "best practices" were affirmed, and a standard was set for what in today's jargon would be called the best available control technology (BACT). However, basic research on air pollution from smelters was put on the back burner for more than a generation. For their day, SCS and the management style that went with it were innovative, even though in retrospect the lack of interest in the effects of the long-range transport of pollution from tall stacks is striking. (Emissions had to go somewhere, and in the 1980s it was discovered that sulfates were being transported to the Seattle-Vancouver metroplex, among other places, far from the stacks at Trail.[24]) To be sure, in the United States this development only became a concern in the early 1970s as the issue of acid rain emerged.

A second lesson is derived from explaining why Judge Hostie and his two colleagues on the arbitration tribunal confined themselves to a carefully circumscribed interpretation of their charge, and why, after voluminous testimony, the tribunal's final decision was so narrowly drawn. Here was the result: in the evolving interdependent relationship between Canada and the United States, both governments came to the conclusion that it was expedient to regard Trail as a special, discrete case rather than as an opportunity to set broad standards or to break new ground. This posture was achieved after both of the parties had used linkage politics to achieve certain goals but then backed off from the consequences of this approach, which could have disrupted other aspects of the relationship.

The Roosevelt administration did not give the Trail case a high priority, in part because there was little stomach in the depths of the Depression for placing limits on industrial production; instead a great deal more enthusiasm existed for job-creating public works projects such as the Grand Coulee Dam on the Columbia River. Providing cheap power and irrigation for farmers took precedence over indemnifying and protecting a few hundred poor settlers in the blighted Northport area. In the politics of regional trade-offs, water projects were a winning issue for Democrats, something that could ally a Carl Hayden, the pro-mining congressman from Arizona, with a populist like the powerful Senator Borah from Idaho. Roosevelt needed these western votes for other programs, such as reciprocal free trade agreements being negotiated with Canada, Mexico, Cuba, and Brazil. As well, the conservationist movement had not yet evolved an agenda or a constituency for protecting the environment from industrial pollution. Smelter cases, even one as important as Trail, could still be looked at as source-associated local problems rather than as part of a regional problem. Nonetheless, the issues could have been broadened as both of the parties resorted to linking issues in their search for leverage to end the long-running Trail dispute on terms favorable to their side.

A project of great interest to the Canadians was the St. Lawrence Seaway, but in 1934 this project failed to achieve support in the U.S. Senate, and a treaty was not signed until the 1950s. D. H. Dinwoodie, a Canadian historian, shows that for a brief period in 1934, Washington linked support for the seaway, a prime Canadian objective, with resolving the Trail dispute.[25] However, this linkage appears to have been a tactical shift away from the earlier policy of doing nothing to interfere with the seaway negotiations. In the waning months of the Hoover administration, the State Department's unwillingness to take a position on the IJC report, which had outraged the Northport farmers expecting a much larger settlement, was frustrating to Jacob Metzger, the hard-driving U.S. agent, who attributed it to the department's concern for the seaway and, as he told colleagues on the Smelter Fumes Committee, a desire "not to hurt anybody's feelings instead of taking a positive stand upon matters."[26] Soon pressure from the farmers and the Washington State congressional delegation led the U.S. government to reject the IJC report and the settlement. The case eventually went to arbitration. Metzger expected to win a much larger award for damage and on terms that would guarantee that pollution from Trail into U.S. territory would cease. For their part, the Smelter Fumes Committee hoped to establish a new air pollution standard for smelters.

In the midst of lengthy negotiations over the scope and structure of the new tribunal, John Read, the astute Canadian agent, engaged in some adroit linkage politics of his own to limit its charge. The idea was to draw attention to cross-border pollution going the other way, especially from Detroit on the American side into Ontario. To this end Canadian scientists from the National Research Council were instructed to obtain the SO_2 air transport data in the Windsor area. Armed with portable recorders, a three-man team led by Morris Katz produced the desired results in March and April, 1934. "We are making every effort to find high concentration of SO_2," he reported, "and have every reason to believe that we will not miss any opportunity which may occur," He noted that the coal-fired Detroit Edison plant was a particularly heavy polluter into Canada, which was indeed a particularly good case: 90 percent of the air pollution in the Detroit-Windsor area came from the American side.[27]

In a note to officials in Washington, External Affairs drew attention "to similar SO_2 visitations in the Windsor area from factories in Detroit. The maximum noted in these tests was twice as large as the maximum recorded in the State of Washington and eight times as great as the maximum which it was proposed in a previous U.S. note should be permitted in the case of the Trail Smelter in future." Although Griffin thought these data could be disputed on grounds that pollution from Canadian industry and emissions from coal-burning home furnaces on both sides of the border could have caused a good deal of it, the survey was useful to Read in his effort to get the United States to back off from its position that *all* harmful emissions from Trail must stop. "The acceptance of the principle of absolute cessation of damage might have shut down the Trail smelter," Read recalled, "but it would also have brought Detroit, Buffalo and Niagara

Falls to an untimely end."[28] For its part, C.M.&S. Company insisted that no standard be recognized that could lead to shutting down the plant. Prime Minister Bennett informed the Americans that the vital smelter industry should not be threatened by the prohibition of "slight and occasional" injury resulting from "uncontrollable circumstances."[29]

From internal evidence it seems clear that Metzger, the head lawyer on the U.S. side, chose to finesse the Detroit-Windsor linkage by maintaining that there was not any general principle being made in the Trail case that was relevant to the Detroit-Windsor situation, which was separate and distinct. And no wonder, Crowe wrote to Read, because the Detroit-Windsor linkage worked to undercut the yardstick argument. The essence of Trail was always "what damage has been done." Other Canadian executives wrote in a similar vein, that is, that the investigation should be confined to the one particular case.[30]

Furthermore, both parties knew it would be undesirable if the litigation should lead to the shutting down of industries throughout North America. The Canadians raised this specter to good result. And yet, however much the parties attempted to limit the investigation to Trail, "there is a particular danger of a rule being established that would inevitably spread to all industries both in their international and national aspects." And if it came to this end, the investigation could then be portrayed, if necessary, as an attack on Canadian industry in general.[31] In any event, both governments were prepared to limit the scope of the investigation so that it would *not* be continentalized. The Canadians and the smelters wanted such an outcome; given other more pressing priorities, the United States could live with it.

A third lesson emerges from a close reading of this history: the smelters acted continentally, across national borders. Only recently have we come to realize the significance of the fact that certain businesses were thinking and acting along continental lines, and such thinking and acting occurred well before the passage of NAFTA.[32] Students of North American business should pay more attention to these earlier networks of executives in the transportation and natural resource sectors, men whose careers in railroads, mining, smelting, electric power, oil, and gas developed a transnational point of view with respect to sharing technology and expertise.

Recall that Consolidated Mining and Smelting Company, the premier smelting industry in Canada, was organized by Walter Hull Aldridge, an American engineer who married a Canadian woman while managing the mining and metallurgical division of the CPR. Later, as president of the Texas Gulf Sulphur Company, he explained the Trail position on emissions to President Hoover and lobbied Congress on behalf of C.M.&S. and the smelter industry during the Roosevelt administration. In turn, the business structure of the Phelps Dodge Corporation as well as the name and location of its smelter on the Mexican border derived from the organizing ability of a Montrealer, Dr. James Douglas. Douglas's interest in metallurgy had been kindled by an American professor of chemistry at McGill University. His grandson, James Douglas, represented southern Arizona in Congress and later became Roosevelt's first director of the Budget Bureau. Percy Gordon Beckett, another Quebecker, was general manager and vice president of Phelps Dodge from 1917 until his retirement in the late 1940s. His Canadian counterpart at ASARCO, Henry Adelbert Guess, the long-serving vice president, "was probably the premier metals authority in North America, the Guggenheim's right hand for forty years, and confidant of Wall Street financiers." In 1905, Guess managed the Cananea smelter in Mexico and then worked in Missouri lead mines before becoming vice president of ASARCO in 1915. His brother George was a well-known professor of metallurgy at the University of Toronto. P. G. Beckett also worked at Cananea, possibly during the great strike of 1907 that was a precursor to the Mexican Revolution, before going on to Argentina and then joining Phelps Dodge in 1908.[33]

Numerous Canadian engineers went on to notable careers in the United States and Mexico. When one of this group, E. P. Matthewson, became president of the American Institute of Mining and Metallurgy in 1923, someone mistakenly claimed that he was the first Canadian-born person to serve in that capacity. It was soon pointed out that in fact he was the institute's *fourth* Canadian-born president. The American Institute of Mining Engineers held one of its meetings in pre-revolutionary Mexico. Many of these men and their families shared the experience of living in the mining camps of the Rocky Mountain West, although the careers of top managers also could include being based in New York, the financial center.[34]

To be in mining and smelting was almost by definition to have an international career. These executives did not cease to be citizens, but the nature of their business led them to derive perspectives that were regional and continental. While "following the mines" across borders, they developed networks of bankers and engineers with whom they shared technical information and passed on and received investment tips about new mining prospects. For example, the tradition of sharing technology served Phelps Dodge well when it decided to develop extensive copper-bearing ore deposits in eastern Arizona and was able to hire a smelter specialist from Anaconda to design the new reduction works at Morenci in 1937. James Douglas, the company's elder statesman, had long believed that however rival companies might compete, their scientists should work together and interchange their findings.[35]

When it came to researching the effects of smelter fumes on crops,

ASARCO took the lead and shared its results with the other mining companies. In what must have been one of the early grants from industry to support research at Stanford University, starting in 1925 ASARCO provided funding to Professor Robert Swain and his graduate students in the Department of Chemistry. Wary of the pitfalls of doing proprietary research, Stanford's president, before putting this ASARCO fellowship through the university's board of trustees, asked Swain to be sure that "there is no question but that the results of any such research will be made available to the scientific public in the usual way."[36] In a long and distinguished career, Swain consulted frequently (including work for Trail), and his students found jobs in the industry.

And so, when the Trail case arose, industry acted within a matrix of relationships, experiences, and procedures that was already well in place. In Metzger's words, "The smelters hang together."

Smelter plant managers on the Mexican border were well aware that the outcome of the Trail dispute could affect their ability to pollute freely into Mexico. The industry position was reported in a closely held survey prepared for internal use by the U.S. Bureau of Mines:

If the tribunal decision is handed down in favor of agricultural interests, the U.S. Department of Agriculture may have to defend instead of prosecute. The smelter at Kellogg will be in a vulnerable position along with the Phelps Dodge Corporation smelter at Douglas, Arizona, and the A.S. and R. Co. smelter at El Paso, Texas. Surely the Mexican Government will be no more tolerant of the amount of smoke drifting across the border from these latter two smelters than we are of the amount coming into the United States from British Columbia. If the theory of invisible injury is true in the Northport valley, it is also true in Mexico. A precedent will have been established and charges will be even more difficult to defend than they now are. Also, it may be possible to find too much SO_2 is going from some of our industrial cities into Canada.[37]

Included in this closely held report was a photo showing a plume several miles in length drifting from the Douglas works into Mexico. "If the present suit should go against the Trail smelter, the residents of Mexico would be in a position to make a real issue of the discharge from this American smelter," the report concluded.[38] In fact, Phelps Dodge first began making damage payments to Mexican farmers in the summer of 1937, as will be discussed later. Half a century would pass before the Douglas smelter was eventually shut down—by the state of Arizona and the U.S. Environmental Protection Agency acting under the authority of the Clean Air Act of 1970.

What is important to note here is that the USBM reflected the views of a process industry it was mandated to serve. Alarmed that the invisible injury thesis might prevail, industry enlisted the USBM to coordinate a comprehensive defense and offered to underwrite the costs, and in late 1936 a program was authorized by Reginald Dean. (For more discussion, see Chapter 4.) USBM field staff began to gather and compare closely held emissions data from each smelter while preparing an inventory and evaluation of all the SO_2 research under way by the smelters and in private laboratories. At Metzger's urging, this sponsored research effort was stopped abruptly by the State Department in July 1936, and permission for a USBM official to attend a strategy session of the smelters in New Brunswick was denied.

Seeing this collaboration between the mine and smelter fraternity and a U.S. government agency as subversive to American interests, the State Department advised that immunity from the Logan Act could not be assured if the smelters persisted in supporting Trail, a foreign corporation. The department had already refused to drop the case or to adjust the U.S. position after consulting with industry. Nor would it allow industry to examine the government's evidence or to divulge the U.S. negotiating position, even though they already had a copy, courtesy of Consolidated. Pressure to fire Metzger on grounds of animus toward the industry was resisted.[39] When industry asked to be a party to the Canadian side of the dispute, this request was denied. Consent to place at the disposal of Consolidated the results of their experiments on invisible injury was also denied.[40]

The Canadian government then took the unusual step of requesting that Americans testifying on behalf of Consolidated be given blanket exemption from the Logan Act. As the U.S. agent in charge, Metzger was outraged at this close linkage with the smelters. Should the Canadians now threaten to refuse to proceed with the arbitration, Metzger advised, they should "be informed that the Government of the United States does not view with favor the action of the Canadian Government in supporting Canadian private interests and United States private interests in their efforts to defeat this arbitration."[41]

It happened that the Logan Act, an obscure antisedition statute dating from the 1790s, was not invoked. But in response to the threat, U.S. smelters did become somewhat more guarded in their attempts to steer or block the scientific investigation, and the USBM withdrew from the agreement to coordinate and conduct research on behalf of industry. Individual Americans did testify at the tribunal hearings, as for example George Hill, the ASARCO agricultural expert. American companies kept in touch with the NRC, witness a letter from Freeport Sulphur Company reporting that as a result of the effort started in the fall of 1936 to obtain

scientific information on the invisible injury thesis, it had been supporting research through a fellowship at Cornell University, whose findings
"should lend little comfort to proponents of [this] hypothesis."[42] There
can be no doubt that Trail managers knew they were carrying the spear
for industry. Read himself was in frequent touch with various U.S.
smelters. Given the speed with which the U.S. scientific case collapsed in
1937, the smelters worried unnecessarily that Consolidated and External
Affairs might not have the skill or strength to kill the invisible injury thesis. Swain's appointment as the Canadians' technical adviser to the tribunal, a move strongly supported by the U.S. smelters, helped in this
regard. Let us review the stakes as seen by industry.

Sulfur producers from Texas and Louisiana laid out their case in February 1936, while W. H. Aldridge, the founder of modern Trail and then
president of the Texas Gulf Sulfur Company, was encouraged by Trail
management to lobby Washington behind the scenes. For the United
States to press on with the invisible injury approach was not helpful to
the sulfur industry, their lawyers told Metzger. According to Metzger,
their worry was that the Canadian interests might not make an adequate
defense against that theory. "I told them," Metzger noted in a memo
recording these pressures, "that they could rely on the Trail Smelter making the best defense that it would be possible to make as the Consolidated
had unlimited resources, was supported by the Canadian Government,
and was assisted by the American Smelting and Refining Company." Although pleasant enough, Metzger saw the interview with the lawyers as
a threat.[43] But what most worried the sulfur producers was that the investigation might turn up new data on acidic soil conditions, indicating
that sulfur was toxic to soils.

In addition, Texas senators writing the secretary on behalf of the sulfur industry pointed out that the dispute had already forced Consolidated to spend $10 million on facilities to control emissions. This
requirement resulted in what they called a large new supply of "waste-
fully produced by-product elemental sulphur which will enter into uneconomic competition with the sulphur mines of the United States," with
fertilizer companies through the superphosphate produced, and with
American nitrogen producers through the ammonium sulphate produced. In fact, they concluded, "this undesired and uneconomic production will be the result of the previous representation by the government
of the United States in behalf of the local claimants in the State of Washington." If the invisible injury thesis should be accepted in claims for
damages, instead of "the commonly accepted visible evidences thereof,
far more Americans will be affected than will be benefitted in the pending arbitration." Whole industries dependent on sulfur-bearing coal
would be put at risk. Thus it was necessary "that due consideration be

given to a proper balancing of the advantages and disadvantages which would result from these questions."[44]

Metzger rejected these arguments point by point, observing that high freight rates would probably limit the competition from Trail fertilizer, and that the complaints made against the principle of invisible injury were very similar to those made to him by an ASARCO attorney several months ago. Interestingly enough, Metzger brushed off the senators' reference to the effect of a ruling on the operations of other industrial plants near international boundaries. In their letter, he said,

> they refer specifically to smelters at El Paso, Texas and the industrial centers of Buffalo and Detroit. Our abandonment of the Trail Smelter arbitration which is suggested in the letter would not relieve us from the necessity of dealing with any questions that arose at El Paso which the Mexican Government brought to our attention. Neither would we be relieved from the necessity of dealing with the situation at Detroit and Buffalo by withdrawing from the present arbitration. The situation at Detroit and Buffalo has been raised by the Canadian Government and they have been answered. Canadians are not lax in pressing grievances that they have against us nor is there any reason to expect that the Mexican Government would hesitate to complain if the smelter at El Paso seriously damaged property in Mexico. All of these matters, if and when they arise, should and doubtless will be dealt with on their merits.[45]

Now it was Metzger who, in his desire to win this particular case, was backing off from linkage and downplaying the larger implications being raised by industry.

The senators' letter had obviously been written for them, according to Metzger, "indicating rather clearly that the sulphur industry, like the United States smelting industry, is cooperating with the Trail Smelter . . . , [and that] since the Convention has been concluded [in 1935 these industries] have sought to suppress the arbitration." Moreover, he concluded, showing his populist colors, if the case should go against them, any curtailment of the sulfur industry would be spread over a large number of American citizens. Were the Stevens County farmers to lose, "their burden would be overwhelming [because] they, in all probability would be deprived of home and livelihood."

Finally, in this long review of the industry position, it is clear that Metzger was confident in the U.S. scientific case: "The Trail Smelter and the American Smelting and Refining Company have betrayed the keenest anxiety in regard to the outcome of this case from the beginning. They use the expression 'invisible injury' to express an idea which we express by 'accumulative and residual effects.' The smelter term was obviously chosen to strengthen their case for visible injury." Here is an indication of how Metzger, had he lived, would have framed the argument. Revealingly, in his fixation with ASARCO as a major mover in the Canadian case, he made no mention of the NRC researchers who, as we have seen, were both well prepared and well supported by their government.

However, in the jockeying and maneuvering leading up to the early and, as it turned out, decisive hearings that transpired in Spokane in July 1937, the focus on ASARCO was by no means misplaced. The context is documented in an extraordinary letter addressed to Secretary of State Cordell Hull from Chandler P. Anderson, an international lawyer with much experience in boundary claims and settlements then lobbying for ASARCO and speaking on behalf of the American smelting companies. Collectively, Anderson noted, his clients operated nearly all the copper and lead smelters in the United States: "The serious concern of these companies is that as a basis for compensation an attempt may be made to prove a theory of loss or injury which, if upheld and adopted as a precedent in this country, would lead, not only to the award of speculative and imaginary damages in this class of cases, but to decrees of injunction which might cripple the smelting industry." If the farmers' claims were upheld by the tribunal as well founded,

> these claims of injury, beyond those for visible, tangible and measurable damage . . . would undoubtedly be used as a precedent to justify similar claims by American agriculturalists against American smelting companies, and, indeed, even claims by Mexican or Canadian agriculturalists against American smelters and other smoke-producing industries along international boundaries. [Such an] international precedent . . . would be destructive to widespread American interests much more important in their scope and value for the general welfare than any possible intangible damage which could be suffered by the very limited number of agriculturalists whose interests could be affected by the operation of the smelters.

In short, a finding that validated damage claims based on the small amounts of SO_2 that are found around all industrial centers would open the entire field to unlimited litigation. "So the results determined by this tribunal may be far-reaching as affecting industrial operations throughout the United States." These industries deserve protection from their government. With this situation in mind, Anderson concluded, the government should affirm that any finding along these lines would not be construed as establishing a general precedent. Furthermore, the jurists appointed to the three-person panel must be free of any association with the present controversy, and in designating a scientist on the part of the United States to assist the tribunal, "the experience and impartiality of the person selected shall be the determining factor."[46]

Given industry's insistence on keeping the dispute in tight bounds, it is little wonder that the tribunal—reflecting also the wishes of the two governments—was loath to push in the direction of establishing a new continental standard for adjudicating pollution disputes. In any event, Dean, the technologist, and Swain, the mainstream ("pro-smelter") scientist, were designated by the Americans and the Canadians, respectively, to advise the commission, which then relied heavily on their counsel.

Before all of this became clear in the summer of 1937, the smelter and sulfur interests were leaguing together, with the coal people probably adding their weight as well, all hoping to mount a coordinated defense under the auspices of the U.S. Bureau of Mines, in confrontation with their own government's position in the dispute. But after Metzger's death the fight went out of the Roosevelt administration, which had never invested much, if any, political capital in Trail anyway. The Department of Agriculture, still pushing a controversial theory, was left to twist in the wind.

Assessing Trail also requires paying attention to the way the two different political systems handled this complex dispute. In Read's words, "The Company and the government were yoked together as a team. . . . [This] link between the Company and the government was maintained at all stages of the procedure."[47] Also, the Canadians had the resources of an entire industry behind them. Field research by NRC scientists was closely coordinated with this winning strategy. A nation one-tenth the size of its neighbor, and only recently in charge of its own foreign policy, marshaled all its resources—intellectual, monetary, organizational, and political—through the concentrated power of its parliamentary system.

In contrast, the U.S. system of divided government was less able to coordinate its resources or even to agree on a coherent strategy. From the Canadian perspective, it was ever so. (See, for example, Canadian accounts of negotiating the Free Trade Agreement in the late 1980s.[48]) As seen from the U.S. perspective, the very responsiveness of the system to grassroots input set a high threshold of expectations for government. In the Trail case, seeking just and fair redress for the farmers of Stevens County in the upper Columbia River Valley was the proximate goal energizing the American response. Metzger and the USDA responded by identifying emotionally with the locals and their plight. However, the American political system was also responsive to pressure from lobbyists, so that elected officials gave conflicting signals to civil servants carrying out U.S. policy. As well, in the U.S. adversarial system, federal agencies often compete against each other for the policy lead; witness the strained relationship between Agriculture and the Bureau of Mines, seeing its mission to serve industry and solve technical problems and disparaging the work of scientists from the same government investigating the chemical effects of air pollution.

Given the asymmetries of the already complex binational relationship, Canadians perforce celebrated their victory in a low-key manner. Read reported on "the rather humiliating collapse of the United States case" in a letter written during the hearings in Spokane:

They have presented reports which are incapable of justification from the scientific point of view and which, if accepted, by inference support a thoroughly unsound position. On the other hand, I feel that the Government, bearing in mind our general relations with our neighbours, cannot afford a position in which U.S. Government scientists are humiliated.[49]

He might have added that sufficient doubts had been raised already in the minds of the arbiters so that further rough handling of scientists by an experienced corporate lawyer would not be necessary.

Although the U.S. side did better in advancing the validity of its scientific evidence in later hearings—with Sherley gone and John Raftis managing the U.S. case—it failed to convince the three arbiters of the invisible injury theory. "By far the most difficult part of the case consisted in issues of fact, highly technical in character, and dependent for their solution on scientific experiment and testimony," Read recalled. To sort out the evidence, the tribunal relied heavily on the scientific judgment of Reginald Dean and Robert Swain, the technical consultants, both of whom as it happened were pro-smelter.[50] In close coordination with the Canadian Meteorological Service, Swain developed the innovative dispersion regime. Dean in particular did not mince words about the quality of work done by USDA scientists. After 1941, both consulted for Consolidated.

The United States was unable to make any case at all to support damages in dollars and cents, and in the face of conflicting scientific testimony, the tribunal awarded only $78,000 in damages for two big burns causing visible marks in the spring of 1934 and 1936. Consolidated also would pay for the experimental regime limiting emissions to some 150 tons per day (100 tons per day during the high growing season) and the salaries of Dean and Swain to administer it. The tedious job of sorting out the claims and distributing this award, and the earlier $350,000 damage payment to the thoroughly discouraged farmers fell to Skinner, who was appointed special master. "It serves him right for all the trouble he caused," Crowe remarked.[51]

The final settlement of damages awarded in April 1938 was a victory for Consolidated, and as Crowe wrote Dean, "We're very pleased indeed with the terms of the award and I have heard from some of the American smelting interests to the same effect," particularly from ASARCO (which had been especially worried about the effects of an unfavorable ruling on its plant operations in El Paso, Texas).[52] Dean and Swain told Crowe that "had it not been for the two burns . . . [we] doubted that any money would have been awarded. The theory of invisible injury advanced by the U.S. they say—and the award indicates the same—was entirely rejected." Furthermore, the fume control regime was not expected to be onerous; rather, it would validate what the Trail smelter was already doing:

The consultants believe that it takes one part per million for at least an hour with humidity of 50% or over before markings occur. . . . This, of course, is our real victory since the U.S. Government for a good many years has been insisting upon a limitation of concentrations to one and two tenths, as you know. There have not been recorded at Columbia Gardens [on the border] since our recorder was installed a year ago, a concentration where one part prevailed for an hour.[53]

On the basis of subsequent experiments with wind currents and plume diffusion patterns, Dean and Swain recommended a more rigorous regime, which the tribunal adopted before it disbanded in 1941.

The investigations begun in 1927 were broad and comprehensive for the times, but it is well to ask what else might have been done differently, particularly with respect to health and investigating other sources of pollution such as lead poisoning and slag dumping in the Columbia River. In particular, the absence of a health component in both the IJC and the tribunal reports is striking, given that the goal of protecting human health became the keystone of the Clean Air Act of 1970. To be sure, the work of Gustave Freeman and others showing the effects on the lungs of pollution in an industrial setting and in particular documenting the effects of ordinary, everyday exposures was decades away.[54]

Health effects in the form of coughing, red eyes, and increased discomfort, particularly in people with lung problems, were noticed immediately when as a result of increased production and the construction of the two taller stacks in 1925 and 1927, sulfur fumes penetrated much deeper into U.S. territory. Years later Marian Bleeker, who worked on the investigation as an assistant to a U.S. soil expert, recalled "hot nights when the stench hung in the air and days when the emissions left the whole area in a noxious fog."[55] In 1928 the newly organized Farmers' Protective Association heard from a D. O. Lewis, a consultant to the Canadian farmers in their earlier dispute with Trail, and other smelter cases, who told them of the health effects from arsenic in the smelter smoke. Yet health was not included in the original 1927 work plan elaborated by the USDA, which was otherwise a full investigation of soil, crop damage, atmospheric conditions, and plant pathology, including controlled experiments to determine the extent of the fumes injury. At the settlers' insistence, health was added in 1928, and $15,000 was appropriated for an investigation by the U.S. Public Health Service, although there was a year's delay in getting it approved by the Budget Bureau. Claims for health damage were included in the opening brief submitted on behalf of the farmers as the IJC began its investigations.[56]

At the State Department's request, the Public Health Service made a feasibility study of the Northport area in August 1930. However, the full investigation never got off the ground, because the Public Health Service would not recommend conducting a short-term investigation on so few people—only 447 children and some 189 families in the affected area—where the effects of SO_2 would be difficult to sort out from the normal expected incidence of respiratory illness of all kinds, including "influenza, tonsillitis, and several other conditions which would not be expected to be associated with any possible hazard from sulphur dioxide." A protracted study over five to ten years might reveal injury to health, but a one- or two-year experiment would be of no value. Consequently, the funds were reassigned to other aspects of the fumes investigation.[57]

In their independent report to the IJC, the two deans (Miller and Howes) had regretted that no experts had been sent to look into this phase of the controversy. "It should be surely as important as the matter of health of animals," they wrote, "although we grant that the possibility of complications is greater and the difficulty of diagnosis is very much greater." Local doctors suspected that certain health problems were related to breathing poison fumes or to eating vegetables exposed to the fumes. "In our opinion," they concluded, "the chief injury is found in suggestion and if, as is probably the case, there is no injury from SO_2, the fact should be made known by independent physicians, or if the contrary be the case, it cannot be known too soon."[58]

The German literature referenced in USDA reports was all published before 1925 and dealt with the effects of air pollution on vegetation. However, in German articles published between 1926 and 1935 on the effects of air pollution, five of the ten papers listed were on damage to human health. Skinner and his USDA colleagues, even the USDA chemists and plant pathologists working in the field, should have consulted this more recent German literature, which indicated that health, as it were, was in the air. In fact, the U.S. Public Health Service embarked on a nationwide study of the extent of urban air pollution and in the early 1930s started to survey the smoke, soot, and sulfur dioxide concentrations in the air of fourteen American cities.[59] Perhaps, in the depths of the Depression, this was another reason why the Public Health Service had no interest in a small town like Northport.

However, in their search for immediate evidence to support claims of damage, Metzger and Skinner never developed much enthusiasm for a health investigation. And because the United States was not giving this matter any serious attention, the Canadians did not spend any time on it either. In the first half of 1929 it was decided to check for lead and arsenic deposition in the air, but all along Skinner thought that this investigation would prove negative. Griffin wrote that atmospheric sampling for lead traces had just been completed, and the preliminary work indicated the results would be negative. The work had been done "very satisfactorily and should definitely settle the phase of the problem under considera-tion, as the fumes came down quite heavily on a number of occasions during the course of this work." Laboratory analysis of samples filtered from air at the Murawski place near the boundary continued into 1930 but turned up negative.[60]

That area soils had been exposed to emissions from the old Northport lead smelter was of course known to everyone. Trail smelter workers had been exposed to dangerous levels of lead until the particulate recovery system at the lead plant was improved in the early 1920s. What was not known until the experimental work of Clair C. Patterson in the 1950s was that ingesting even small amounts of lead was a dire health threat, to chil-dren particularly. Patterson's research was cited by environmentalists and scientists who successfully lobbied for the enactment of the Clean Air Act of 1970 and the phasing out of lead in gasoline, the main source of lead in the atmosphere.[61] Given the knowledge then available, it is highly un-likely that this investigation on the possible health effects from airborne lead deposition could have turned up any new results.

Griffin then stumbled on another source of heavy metals pollution in the course of testing samples of Columbia River water for Ph values, to-tal solids, sulfates, chlorides, arsenic, zinc, and lead. Residents were com-plaining that a progressive diminution of the fish and wildfowl populations had occurred in the past ten years, that horses grazing on low banks in the Northport region had sickened, and that persons were becoming ill after drinking the water. In a series of tests during 1932 and 1933, Griffin "found that vast quantities of slag have been brought down the Columbia river . . . and some of these specimens were found to con-tain appreciable amounts of lead and zinc." It was his opinion that lead blast-furnace slag was finding its way into the river "in large quantities thus constituting a question of water pollution of considerable impor-tance." A chart was prepared on the water quality of the river and some tributary streams for the period June 28, 1932, to February 3, 1933.[62] He then prepared an eighty-five-page paper on the subject, which concluded by recommending that all smelter wastes be excluded from the Columbia River and in particular that slag pollution from Trail be abated.

Although no further field research was done, Griffin was onto something. However, the American attempt to secure a ruling on these potentially very interesting findings was disallowed by the tribunal on the grounds that damage from this source (slag) had not been established. For their part the Canadians maintained that the question of polluted waters fell outside the terms of reference, which was to investigate the U.S. charge that "fumes discharged from the Smelter . . . have been causing damage in the State of Washington." As Crowe put it, "This subject is [therefore] irrelevant to the present case."[63] Cautious to a fault, the judges let slip an important line of inquiry. By the time the practice was halted in 1991, after nearly a century of dumping, an estimated fifteen million tons of slag chunks and particles were mixed in sandbars along the banks of the Columbia south of Trail or had been carried into the depths of Lake Roosevelt.[64]

Rebuffed by the tribunal, Griffin still wanted to revisit the subject of slag and water pollution and to publish the results, which he thought would be of general interest since the Grand Coulee Dam project was then under way. Referring once again to this earlier work, he informed Skinner that only one example of what he called "gross industrial river pollution" had been found: the deposition of finely divided, insoluble, lead blast-furnace slag, some of which had been carried far downstream into the United States and was found in shifting bars in the water and lodged along the river's edge. Knowing more about the chemical properties of the water "would be of interest to the agriculturalist using irrigation water from the dam, and to the engineer as well as to the chemist." This information was important, because water impounded in Lake Roosevelt and power from the dam were slated to support a large irrigation project to benefit small farmers, a New Deal project that largely failed.[65] The slag study was one of three papers Griffin proposed to publish out of the Smelter Fumes Committee's work, the others being the effects of low-level acid and sulfate contamination of rain and snow and the corrosive effects of smelter smoke on iron and steel. Nothing came of it, and the slag deposition work was yet another of the leads coming out of this research that was soon lost to science and then to history. Washington State chemists investigating slag pollution in the late 1980s did not reference this early work.

To conclude this case study of the Trail smelter dispute, the last word, fittingly enough, comes from the Canadian brief presented to the IJC in 1930. Here is glimpsed the regional problem to come, requiring new measures to protect what a later generation would call the continental airshed. In 1930 these facts were apparent: not only was this the first time a transborder nuisance of this sort had come before an international tribunal, but also it was almost inevitable with growing industrialization that there would be more questions such as these between the United States, Canada, and other countries. "Questions may arise between the United States and Mexico; questions may arise between the United States and Canada in respect to either of the two main boundary lines"; today it was a U.S. claim, tomorrow it could well be reversed.[66]

Seventy years later it is probable that there will be a Clean Air Agreement for North America. (Thoughts on the need for a continental regime, and the Transboundary Environmental Impact Assessment that would accompany it, are presented in the last chapter of this book.) That seventy years ago both sides in the Trail dispute shrank from the wider implications of the dispute, limiting the findings to this one case, indicates just how far we have come in recent years toward recognizing, and acting for, the North American commons.

PART II: THE GRAY TRIANGLE
CONFRONTATIONS OF THE 1980s

If the protracted Trail smelter dispute on balance favored the North American smelter industry, forty years later the dispute over the Douglas smelter in what came to be called the Gray Triangle on the Arizona-Sonora border was an unambiguous victory for tighter air standards. In the first case, an innovative control system (SCS) was developed in conjunction with other measures to alleviate the welfare concerns of local people while still accommodating the venting of substantial, if reduced, amounts of sulfates into the air. In the Douglas case, the effects of air pollution on environmental health and on nature itself were the central issues, and producers were subjected to a higher standard of capture than was imposed on Trail by the Arbitral Tribunal. Once more, local people played important roles, but this time the institutions and the conjuncture were more favorable to their interests. And the issue of transborder air pollution, recognized but finessed in the Trail settlement, was both recognized and addressed in the 1983 La Paz agreement between the United States and Mexico. In North America, La Paz was the first transnational air agreement, antedating the 1991 acid rain agreement between Canada and the United States by several years.

To be sure, abatement was not a zero-sum game for industry, because better control of stack gases and fugitive emissions generally meant more efficient production techniques as well as better health and welfare for local people. However, the "tilt" favoring one or the other constituency was important. Achieving balance among the interested parties was the role of politics. Achieving a more lasting benefit to North American industry and society as a whole was the aim of public policy.

What is new in this second case study is the enhanced status and authority of science in the law under the 1970 Clean Air Act, as well as the critical role of the U.S. Environmental Protection Agency (EPA), founded in 1971, and the rapid growth of environmental nongovernmental organizations known as ENGOS. Driving and sustaining these new actors was what political scientist Lynton Keith Caldwell has called "the emergence of an ecological perspective" that "has become widely shared among better informed, better educated, and more affluent Canadians and Americans," and would later be embraced by their Mexican counterparts as well. "In its purest form," he wrote, "the ecological perspective clashes with conventional political and economic values across a broad range of specifics involving attitudes toward economic growth, legal rights (particularly in the ownership and use of property), national sovereignty, and the necessities of national security." Moreover, he continues, "The ecological perspective on life and the world for North Americans tends to be continental in a global context."[1]

These strands came together, sustaining a much bigger tent, making it easier for interested parties to make contact and connect. With the onset of regionalization, the institutions finally started to catch up with the mining and smelter industry, which decades before had developed a strategic conception of North America.

Producers had to cope with the uncertainties of price fluctuations in the international commodities markets and, once controls (or the prospect of controls) of emissions came in, with the certainty of having to dispose of the unwanted by-products—sulfur and sulfuric acid, for which there often was no ready market. Finding a business solution to the nasty aspects of smelter pollution was difficult. Under the innovative leadership of Selwyn Blaylock, Trail had found a viable business solution in its Elephant brand superphosphate fertilizer, but Arizona smelters were not well placed to market low-value by-products. So they captured the unwanted gases when they had to but preferred to vent them for as long as possible, using SCS as long as possible. The Douglas smelter is the classic case, but this delaying tactic was a rear-guard action.

Location in scarcely populated, arid regions like the American Southwest could not overcome the rising public alarm over the long-range transport of acid rain and the developing scientific and legal conception of the airshed as a whole. Modern flash furnace technology, such as Phelps Dodge installed at Playas, New Mexico, in the 1970s, eliminated the smelter plumes that had once seemed an inevitable part of western landscapes. But this technology was expensive and could be tricky to operate. Temporizing was another business strategy, an alternative to using the best available control technology. Under the shield of corporate law, batteries of lawyers stoked the fires of litigation, keeping science and the greater public interest at bay. Other smelters and other companies were players in this corporate game, but none was as significant for the outcome of the transnational story on the southern border as the Douglas Reduction Works (DRW) of the Phelps Dodge Corporation, located just six miles from the Mexican border.[2]

Shut down in 1987 for noncompliance with the Clean Air Act, the DRW was an old plant, basically using the technology of 1914 when most of it was built. Although the DRW's reverberatory furnaces had been converted in the 1930s to accept natural gas, sulfides from the copper ore were vented freely into the atmosphere. A small acid plant was operated intermittently but ceased to operate during the Depression. Late in the game a supplementary control system to disperse the smoke was installed, which was a defensive move to forestall the installation of modern abatement technology. Ordered by the EPA to rehabilitate the works to modern control standards or close, Phelps Dodge secured temporary operating permits for as long as it could and then elected to shut Douglas down.

Douglas was well located to smelt copper concentrates from company mines in Sonora and nearby Bisbee, but with the sale or closing of these mines in the 1970s Douglas no longer had a place in PD's long-range planning. A modern smelter was built at Playas, New Mexico, in compliance with the Clean Air Act. Yet Douglas, though obsolete, was still profitable to operate without controls. Faced with the threat of government action, the decision was to operate as long as possible without modernizing the plant. The contrast with Trail is striking. Under pressure, the Canadian Pacific Railway subsidiary used the threat of regulation as a spur to technology-forcing. Under pressure, Phelps Dodge fought it out, using Congress, the hearing process, and the courts to delay shutdown. Why was this particular business plan adopted?

Trail and Douglas are landmark cases in the annals of North American air pollution. That the industry acted continentally is a point well established in the earlier chapters. Competitors when it came to mines and markets, the corporations regarded both technology and their research on the effects of SO_2 as fungible parts of their corporate commons. Thus they cooperated across international borders to fight or manage the threat of regulation, and for telling this story the two cases are tailor-made.

With respect to transborder air pollution, Douglas was the smoking gun. A 1936 Bureau of Mines report, with its photo of the Douglas plume drifting into Mexico, concluded: "If the present suit should go against the Trail smelter, the residents of Mexico would be in a position to make a real issue of the discharge from this American smelter."[3] With Canada pushing hard for an acid rain agreement in the 1980s, the Reagan administration had no desire to be drawn into an environmental dispute on the other border, specifically the Gray Triangle of smelters including Douglas in Arizona and Nacozari and Cananea in Sonora. The 1983 La Paz agreement provided a framework for developing a pollution abatement plan with Mexico. Later, with the passage of NAFTA in 1993, the three governments had an institutional umbrella to address problems of the North American airshed as a whole.

PHELPS DODGE OPERATIONS IN ARIZONA – NEW MEXICO – MEXICO:
THE FIRST 100 YEARS

"Phelps Dodge Operations in Arizona—New Mexico—Mexico: The First 100 Years." (*Pay Dirt*, special supplement, Summer 1981, p. 5)

The Douglas smelter in 1936, with smoke going over the border into Mexico. (Ambrose Report, U.S. Bureau of Mines)

To be sure, the smelters had been operating continentally for years, defined as a shared strategic conception of North American space rather than in terms of ownership and controls that had little relevance to environmental issues. Thus, for the purposes of this book, mine and smelter history is a good way to analyze both borders in conjunction. British Columbia–Washington and Arizona-Sonora were constituent as well as interdependent parts of the same mineralized West. The smelters shared information on wage rates and work rules. They shared technology. They shared in the boom-and-bust cycles of a common resource-based economy. And when it came to the issue of pollution controls, "the smelters hang together," as a U.S. offical put it.

British Columbia was the first to break loose from hinterland status—when Trail was built up to process the lead and zinc ores previously smelted in Spokane, and ASARCO sold the Sullivan mine to what became the Consolidated Mining and Smelting Company. Sonora delinked in the 1970s, as Mexico developed its own copper industry on the basis of properties previously owned by Americans and managed from Arizona.

But delinked is too strong a word, given the transfer of technology, flow of funds, and social ties across both borders. Phelps Dodge sold the converter plant and other equipment from Douglas to Cananea, while the new Mexican smelter at Nacozari benefited from the operating experience of ASARCO, which held a minority position, Finnish technology in the furnaces, and Canadian finance and construction engineering in the acid plant. Most recently, smelting at Nacozari has been upgraded again with the purchase of the new converter furnace process developed by Chile's CODELCO. If the city of Douglas lost 350 jobs and a multimillion-dollar payroll when the smelter closed, it was rapidly being transformed

Aerial view of the Douglas smelter (1983?). (Lighthawk photo from Dick Kamp)

The Douglas Smelter, with smoke going into Mexico (1983?). (Lighthawk photo from Dick Kamp)

into a service center for retirees, a prison, and the burgeoning maquiladoras on the Agua Prieta side of the line. Meantime, on the northern border, Northport still languished as the distant also-ran to Trail.

My argument is developed in three chapters. In the first I deal descriptively with the mine and smelter industry, focusing on Douglas. The enforcement role of the EPA and the activist role of the nongovernmental organizations are assessed together in the next chapter. The section then concludes with an assessment of transborder linkages in the so-called Gray Triangle: Douglas, Cananea, and Nacozari.

CHAPTER FOUR

Smoke from Douglas

Gleaming in the desert sun, the copper dome over Arizona's capitol proclaimed the reign of copper at the statehouse—from territorial days, when copper and cattle were the leading wealth producers, until well into the 1960s. To be sure, by the early twentieth century it had become the "four Cs"—copper, cattle, cotton, and citrus—and the majority of Arizona's population was still rural. Then the rapid growth of Phoenix and Tucson owing to postwar in-migration heralded a strikingly different demographic profile. By 1960, the state was nearly 70 percent urban. A new economy based on manufacturing, tourism, and services had edged out mining, which was in a secular decline, and agriculture.

No single industry dominated the state as copper mining, agriculture, and the railroads had done for so long. By 1990, just over one-half percent of Arizona's total employment was in metal mining. In fact, the decade of the 1980s, which commands the great bulk of our attention in this second half of the book, saw mining employment drop by half.[1] Big copper was still potent at the statehouse, but the era when corporations like Phelps Dodge, Kennecott, Magma, and ASARCO could count on the congressional delegation in Washington, acting in alliance with other senators and congressmen from western resource states, to protect their industry's interests was waning even before the bust in commodity prices after the second oil shock in 1979 threw the industry into turmoil. Moreover, just as Arizonans started voting for large sums for education, so they were becoming less tolerant of the air and water pollution that had always gone along with mining.

In the old West, developing natural resources brought people, jobs, and progress to an empty land. Mining was an enhancement, connecting this West to the industrial centers whose products depended on metals from extractive industries. In the new West, the costs of mining were calculated along with the benefits. Pollution, the human costs of boom-and-bust mining cycles, the permanent changes to the land—all were factored into a calculus that before had stressed only the benefits to society of well-paying jobs and the unlocking of nature's wealth. Mining continued, but heroic imagery gave way to defensive press releases. Visitors to the museum now housed in the old Copper Queen division in Bisbee, Arizona, are impressed by the business acumen of the men whose portraits grace the walls, including James Douglas, the Canadian metallurgist

112

from Montreal who built the modern Phelps Dodge Corporation. Just down the road, however, as visitors gaze into the disturbing vastness of the open pit Lavender mine, a typed notice reminds them that the copper products on which America's industry and consumer society were built came from holes like this one. Absent such an explanation, the view through a chain-link fence at the highway's edge would be difficult for outsiders and others to interpret today.

To read this book, then, is to enter into an industry and a culture in transition, caught in the mix of old and new Wests while navigating the seas of globalization. The copper industry had long been geared to international pricing. The 1980s brought challenges from old competitors, such as Chile and Canada, while witnessing the rise of new producers, including Mexico, right next door. This marketing situation is a necessary subplot. The main action concerns the new regulatory climate brought in by the 1970 Clean Air Act, the actors and forces at play, and the Phelps Dodge Corporation's business strategy to survive.

As the decade of the 1980s began, Phelps Dodge was still very visible in the economic and social landscape of the state. Active in copper mining since the 1880s, the company grew by mergers and acquisitions to be the nation's largest copper producer. In 1983, *Fortune* magazine listed Phelps Dodge as America's 300th largest corporation, with $1 billion in annual sales, twice that much in assets, and 9,100 employees. But the reality was that the company was poised to go under. Buffeted by falling copper prices, foreign competition, and tightening environmental regulations, Phelps Dodge survived only by dint of a harsh and drastic restructuring.[2] Although air pollution is the primary topic here, this part of the story would be truncated without some discussion of the business realities of mining and smelting, company culture, and the way of life in western mining towns.

Copper is the main decorative motif in the old Phelps Dodge Mercantile department store serving Bisbee, the longtime headquarters for the Phelps Dodge Corporation. It was the flagship in a chain of company-owned stores where managers and miners and their families purchased the necessities of life. Often the employees lived in company housing. One-industry towns studded Phelps-Dodge country, from the older installations at Bisbee, with its mine, and at Douglas, with its smelter, built on the border at the turn of the century, to the newer mine and smelter complex opened in the late 1930s at Morenci, which was also served by nearby Clifton, an old mining town in southeast Arizona. Other towns in the Phelps Dodge system included Ajo, where the mine and smelter opened in 1950, and Jerome with its smelter in Clarkdale, acquired in the 1930s. In New Mexico, the town of Dawson provided coal for the Southern Pacific Railroad and coke for the Douglas Reduction Works. Hidalgo mine in southwestern New Mexico came on line after the war. And in Mexico, the mine at Nacozari, in which Phelps Dodge held a minority interest with Anaconda, had its own community and sent copper concentrates to Douglas well into the 1970s.

These and other company towns were usually the largest source of jobs, payrolls, and taxes in the host counties, whose economies closely mirrored the mining cycle, based on the boom-and-bust phenomenon associated with international commodity prices. Jobs, which became high paying with unionization, were welcome, but local prosperity was illusory because the ups and downs of mining led to underinvestment in local businesses and services, a feature of mining towns throughout the West.[3] Paternalism was pervasive: the company that owned the town exercised considerable authority over the lives of workers and their families. In the industrial culture of early-twentieth-century America, this relationship was seen as a social good. Phelps Dodge participated in the ethos of progressive company-town planning and was loath to let go in the face of labor militancy. But living with impermanence due to layoffs and a history of bitter strikes and confrontations with management also determined the tenor of town life.

Highly stratified by race and class, the copper mining towns were intensely communitarian all the same. Pride in the local high school teams was intensely felt. For example, Dawson fielded a state champion baseball team, just as in British Columbia the Trail Smoke Eaters competed on the semipro hockey circuit. Bisbee and Douglas were intense rivals in basketball. Living as they did in a contingent world, miners and townsfolk developed a strong sense of place as a way to deal with change. Towns without much of a history held on fiercely to what they had. Eclipsed by Trail, the old mining town of Rossland, British Columbia, "was a familiar tale on the resource frontier: after the roaring days, the long twilight of decline. However fleeting its moment of glory, Rossland was briefly a very important place."[4] To grow up in these isolated mining towns was to be marked for life with an intense connectedness to people and place.

This connection remained even when a place was re-created, as happened at Morenci, where the town was literally consumed by the encroaching open pit during World War II, which were boom years for copper. An earlier town, Metcalf, disappeared, but the name continues in the huge open pit mine where the town once was. Of Lowell, a small town close to Bisbee, all that remains is one block of stores perched on the edge of Lavender Pit, a huge open pit mine started in the 1950s to access the low-grade copper from the Bisbee East ore body. Clifton, once a smelter town in its own right until the plant shut in 1932, was hard hit by the severe Morenci strike of the early 1980s along with flooding on the

San Francisco River that wiped out Chase Creek, with its historic buildings and Hispanic housing. But the town endures.

Even when a company town died, the sense of place lingered on. A community of six thousand people simply vanished when the Stag Canyon division at Dawson shut down in 1950, yet every two years Dawsonites return to their bleak former township with its empty foundations and abandoned coke ovens, recapturing for a few hours their lost local world.[5] Of the PD hospital (where I was born) all that remains is a concrete staircase leading upward to a big empty sky. A long strike and the Southern Pacific's conversion to diesel were factors in the decision to close the mine, but the switch of Douglas smelter to natural gas in 1936 meant that coking coal was no longer needed in the PD system, which sealed the fate of Dawson.

A similar fate befell Pilares de Nacozari, the purpose-built town located near the mine in northeast Sonora that provided concentrates to Douglas from the early 1900s until 1931 and again from 1937 to 1949 when the underground mine shut down for good. In this case, exhaustion of the ore reserves was the reason for closing, with up to fifteen hundred workers laid off. The entire facility of the Moctezuma Copper Company division was abandoned, although some leach operations continued until 1960. Later, the property was bought by Mexicana de Cobre, which built a modern flash-furnace smelter at La Caridad, its open pit facility at Nacozari.

Jerome, with its Clarkdale smelter, was acquired in 1935 from United Verde Copper Company, which had opened the mine in 1915. Hundreds of miners and their families, many of Central European origin, spent their working lives there until the mine gave out in the early 1950s. Those who stayed behind in Jerome staved off its fate of becoming another ghost town by marketing their picturesque community as a destination for artists, retirees, and tourists. Then the mine revived again in the 1980s but with an important difference: the old union work rules had been replaced by more efficient manning practices. Productivity increases in open pit mines were accomplished by reducing labor costs and switching from railcars to huge trucks, which meant fewer jobs and fewer townspeople for Jerome.

Bisbee became a shell of its former self when the open pit Lavender mine ceased production in the late 1970s, and headquarters for the Western Division moved to Douglas and then to Phoenix. Although the built environment remains, the bars, shops, schools, churches, and YMCA serve many fewer people these days. Leach mining begun in the late 1980s has maintained some copper production. But the economy now is based on tourism, crafts, and services catering to earnest seekers and dropouts from other parts of the United States.

Laid out by James Douglas and others in the early 1900s, the town of Douglas was a model of progressive planning in a one-industry community. Unlike Dawson and Ajo, however, housing was privately owned but just as rigidly segregated by hierarchy and ethnicity.[6] Douglas lost 100 managerial and staff positions when PD's western operations were consolidated under the new Phoenix headquarters in 1982, and an additional 350 jobs and a multimillion-dollar payroll when the smelter shut in 1987. But retail sales to Mexicans coming from Agua Prieta, its twin city just across the border, and a new county jail helped keep the town alive. In 1987 the twenty-six maquiladoras (in-bond assembly plants) in Agua Prieta and associated administrative and warehousing jobs on the U.S. side were contributing a combined payroll of $23 million, more than twice the smelter payroll.[7] Moreover, in both Bisbee and Douglas newcomers were migrating in—retirees, service providers, artists, and a cross-section of those fleeing big-city life for a variety of reasons—creating a service economy, businesses, and markets along with their own new sense of place, quite different from the old-timers.

Yet the perception of change comes reluctantly to mining towns. In the twilight years of Douglas smelter and until the plant was shut down on January 15, 1987, many residents still could not conceive of Douglas without the smelter. It was the reason for the town, the hub of a rail network bringing concentrates from Bisbee and Nacozari and coking coal from Dawson. At the founding there were two smelters, one of them managed by the company that became Phelps Dodge. In 1931 these were consolidated into one large smelter, called the Douglas Reduction Works (DRW). With its twin stacks and ever-present plume (except for maintenance shutdowns or when the workers were on strike), the DRW was the most prominent man-made feature on the land.

Modern in its day, the plant used a three-phase reduction process: the roasters that produced the charge, the reverberatory furnaces producing matte copper, and the converters that delivered blister copper, which was then further refined at the plant and later in El Paso. Technologically outmoded by the 1960s, the DRW was nonetheless cheap to operate and profitable. The men were highly skilled in getting value from the old facility, which had long since been capitalized. However, it produced large quantities of gaseous by-products, mostly SO_2, which were vented through the stacks. Direct damage to crops was acknowledged and paid for according to long-established practices in the industry, which was by counting visible leaf burn. As long as the public airshed was considered a free good, the smelter was viable. But when it was classified as an "uncontrolled" plant by the Environmental Protection Agency in the early 1970s, the question became how and when Douglas would become pollution-free.

Historians today are especially interested in the workforce, more so than in the corporate side of mining. There is understandably more sympathy for the underdogs than for the managers and decision-makers who made the business go, and as a result, labor strife is well documented in the history of western copper mining.[8] Militant labor had been a fact of life since the 1900s. As already discussed, company unionism was the strategy of choice at Trail during George Blaylock's long reign, who only ceded ground reluctantly to the CIO during the Second World War. James Douglas, who masterminded Phelps Dodge operations in Arizona for more than forty years, was wedded to what he called "cooperation under corporate control."[9]

As at Trail, this policy included a large quotient of benevolent paternalism and in PD's company-run towns the construction of schools, YMCAs, churches, and workers' housing. However, in 1917 Douglas's son Walter battled the unions head-on in a bitter strike called by the International Workers of the World (the Wobblies), which was broken with the mass deportation of strikers under armed guard. This strike was one of the most notorious labor-management conflicts in the West. Another occurred in Colorado, leading to the famous Ludlow Massacre of coal miners and their families by trigger-happy troops from the National Guard, giving birth to the United Mine Workers of America. Dealing with John L. Lewis's United Mine Workers as manager of labor relations in the immediate postwar years was my Grandfather Davis's last assignment with Phelps Dodge. His final contribution to western mining was to serve as receiver for the bankrupt Victor-American Coal Company at Delagua, Colorado, near the site of the Ludlow affair.

Later, the prolonged and bitter strike of 1983–1985 at Morenci, PD's flagship mine, which ended in defeat for the United Steelworkers, has been seen as an important chapter in the decline of organized labor nationwide. Less well covered is the corporate side of the story: that the strike resulted from the union's unwillingness to give back certain arcane work rules and rigid job classifications that had to go in order for the company to survive the steep plunge in copper prices in the early 1980s. Having acceded to similar demands by United States Steel a few months before, the Steelworkers, hoping to stop a national trend, were unwilling to make similar concessions to Phelps Dodge. Ten years later the union agreed to more flexible work rules at Trail when COMINCO said that the company's future was on the line.[10] But in Arizona their defeat enabled Phelps Dodge to become a nonunion company; productivity went up with fewer workers doing a wider variety of tasks. This outcome was a key element in the massive restructuring program that saved the company. During this round of strikes and downsizing, workers who bolted the union were given considerate treatment in an attempt to improve morale at PD facilities.[11]

Anglo-Hispanic conflict emerged as an important residue of the Morenci strike, when among other things many Hispanic members of the Steelworkers were replaced by nonunion Anglos. Furthermore, the strike hurt the company's image in Arizona. Nursing their grievances, the Steelworkers joined forces with environmentalists, thus broadening the social base for the anti–Phelps Dodge coalition seeking to shut Douglas down. This development meant that environmental activists did not have to deal with an Anglo-Chicano split.[12]

Douglas was now very much a Hispanic town despite the influx of newcomers, while Bisbee maintained Anglo social allegiances. Although Anglo and Hispanic miners shared a long history of labor militancy, they lived in segregated communities and followed a nineteenth-century social code based on class divisions well into the postwar era. Managers and miners lived separate lives in different neighborhoods. Historian Thomas Sheridan points out that "the copper towns may have been ethnic melting pots for Europeans, but the boundaries between Anglos and Mexicans remained strong." Differential wage scales were another reason for the division. Thus the towns were far from being multicultural in the modern sense. The European-born had their own ethnic pecking order, with Cornishmen at the top followed by Greeks, Finns, Yugoslavs and other Central Europeans, and Italians, while the native-born Hispanics and Mexicans moved back and forth across the border in a long-established labor flow.[13] This long history of ethnic division is important in the overall history of mining and smelting, but, as mentioned, it was not a factor in the effort to shut Douglas down in the 1980s. The cycling in and out of Mexican nationals was used by the copper companies to control labor costs and frustrate unionization. Well into the 1920s it was common practice to employ a dual wage scale, one for Anglos, the other for Hispanics. Keeping the border fluid with virtually no enforcement while securing exceptions to the quota on Mexican workers enacted in the 1920s were both part and parcel of the marching orders that the copper companies gave their minions in Washington, D.C.

Senator Carl Hayden, whose years of service representing Arizona in the Congress from 1912 to 1969 spanned more than half a century, was a champion of transborder labor flows to supply Arizona's copper and cotton industries. In the early 1930s, to protect the domestic market from the inroads of Chilean and African copper producers using cheap labor, Hayden became a convert to the protective tariff. Arguing against the U.S. copper tariff, the Cerro del Pasco Corporation, which operated in Peru, claimed that the Michigan and Southern Arizona copper companies were "making a strong appeal to prejudice on the labor question, quite ignoring the large use of Mexican labor in Arizona due to the lack of an immigration quota."[14] To be sure, large numbers of Mexican miners were discharged during the Depression. While sending workers back to Mexico was always an option for management, preserving labor availability was the key.

Keenly interested in border affairs throughout his long career, in the 1930s Hayden also monitored the cattle trade with Canada as well as Mexico, arguing against any linkage of a liberalized flow of cattle to the lowering of copper tariffs on Canadian or Mexican imports.[15] He also backed the ranchers in their desire to quarantine Mexican cattle for reasons of disease control and to protect their markets. Hayden is best known for his championship of big water projects in the West, including what is now regarded as the environmentally unsound Central Arizona Project that brought Colorado River water to farmers and the hordes of newcomers streaming into Phoenix and the Salt River Valley after the war. Still, his long-term concerns for copper and cattle show Hayden was attuned to the transnational aspects of Arizona's economy. From 1927 to 1968, in the years he served on the Senate Appropriations Committee, Hayden's low-key approach to looking after Arizona's interests was most effective.

Hayden voted to reduce the appropriation for scientific research being conducted on the Trail smelter case by the USDA's Smelter Fumes Committee. Furthermore, his interest in water reclamation projects dovetailed with those of other western senators and the Roosevelt administration. These men wanted dams, like the Grand Coulee on the Columbia River with its Lake Roosevelt stretching almost into Canada, not limits on industrial production. In the politics of regional tradeoffs, it is likely that he abetted the declining interest of official Washington in the Trail air pollution dispute.

By 1935 the Senate Appropriations Committee was eager to end this funding for research on Trail smelter pollution. Hayden spoke up for acid recovery as the preferred remedy for the industry as a whole and cited the solution achieved by the Ducktown smelter in Tennessee, which shipped the sulfuric acid to Florida and applied it to phosphate rock. Hayden recommended this same process for Trail, knowing that Consolidated was already taking this route. However, the western smelters were not so fortunate: "The difficulty in many of the Western States, such as Arizona and Nevada, is that phosphate rock cannot be found near enough to our smelters to make what is called acid phosphate for fertilizer. Of course, it is unprofitable to use sulphur smoke to make sulfuric acid where there is no place to sell the acid."[16] In other words, there was not much they could do: the sale of smelter by-products was not an option.

Moreover, there was a new twist: if the United States won a big judgment against Canada, Hayden argued echoing the smelters' position, "It is entirely conceivable [that] the same rule might be a basis for Mexico to lay a claim against the United States. I arrived at that conclusion by seeing the sulphur smoke blowing from [Douglas smelter] into Mexico."[17] And since Douglas and the large ASARCO copper and lead smelter in El Paso might be damaging agricultural lands in Mexico, the subtext of his remarks is that there could be no interest in pursuing the scientific investigation into the subtle effects of SO_2 on vegetation. The invisible injury thesis was an industry killer. Hayden's views were very much the majority position in government, industry, and the scientific community.

Writing in his diary, Henry G. Knight of the Smelter Fumes Committee told a sympathetic senator that "there was considerable tension and the American Smelter Company [ASARCO] would undoubtedly go to almost any lengths to defeat this case against Canada as any decision would have a very far reaching effect and probably would affect the status of smelters along the Mexican border involving American interests."[18] Closely connected to industry, Bureau of Mines officials were unwilling to countenance research that could be seen as antismelter and disparaged the work of their USDA colleagues. That the industry favored a combination of dispersion and marketing solutions to the SO_2 abatement problem is clear, although Trail itself was unique in its commitment to technical innovations, as has already been discussed. What, then, were smelters in the American West to do?

In 1928 Stuart Griffin reported to Skinner on a site visit to the Anaconda smelter, where a large proportion of the sulfuric acid was used to leach copper ore. The company was also marketing superphosphate fertilizer made from the acid. "It seemed," he concluded, "that the Anaconda people had found an economically feasible way in which to use a great deal of sulfuric acid." Blaylock at the Trail smelter told Griffin that with their ores, leaching was not practicable in most cases. His company was seriously interested in superphosphate, although developing a good market for it in western Canada was a slow business.[19] However, interest in leaching waned with the Depression and low copper prices.

Sulfur recovery was simply not an option, various smelter representatives agreed in the early 1930s at a meeting in Salt Lake. "Because of the enormous work necessary to develop a process, the high cost of by-product sulfur as compared to Gulf sulfur, and a limited market because of high freight rates, sulfur recovery would not be feasible as a general solution of smoke damage," a USBM official summarized. It was agreed that only one company could do this profitably. United Verde Company at Clarkdale (acquired by Phelps Dodge in 1935) opted to be the one and was still working on process technology in 1934, when the USBM conducted a survey of the industry. To rid themselves of the unwanted sulfur gas, the others used high stacks and purchased smoke easements.[20]

In 1934 the State Department asked the USBM to conduct a survey of smoke abatement practices among American smelters, which was done. Dispersion of heated gas from tall stacks was the method of choice, given the difficulties of marketing either acid or elemental sulfur. Yet despite the large quantity of sulfur dioxide that was "thrown into the air in the area surrounding any large industrial city" by coal-fired power plants, an amount comparable to even the largest smelters, "no damage has been observed to crops or animals. Dilution is, therefore, a practical method of handling the sulphur smoke problem [for smelters]," the USBM concluded.[21]

Copper ore with its high sulfide content is more difficult to treat than zinc or lead. Thus, with the exception of the plants in Tennessee, "and to a lesser extent the plants at Douglas, Garfield [ASARCO's Utah facility] and Anaconda, the copper smelters in the United States do not fix sulphur dioxide as sulfuric acid." None of the U.S. lead plants recovered acid, but because the percentage of SO_2 involved in zinc roasting is higher and therefore more profitable, most zinc plants, which are also close to markets, did.[22]

Of particular interest to the State Department as it prepared for the second round of negotiations with Canada was knowing the ratio of the amount of sulfur controlled to the amount produced in the main smelter areas of the United States. Of the smelters surveyed in Arizona—Clifton, Miami, Superior, and Hayden—none were capturing SO_2 in 1933, while the recovery ratio of Utah smelters was less than 5 percent.[23] But the data in this 1935 report are remarkably thin. Plant managers were reluctant to divulge air pollution data that might be used against them in litigation. (Under the 1970 Clean Air Act, smelters were required to furnish emissions data for the public record.)

Given their predilection for technical solutions to air pollution, the USBM's Metallurgical Division under R. S. Dean was hostile to the approach being taken by the USDA scientists with their stress on chronic and subtle plant injury. Cooperation between the Bureau and industry increased with the possibility that the U.S. government case against Canada might be sustained, based on new scientific information on the effects of SO_2 on plants and vegetation. A program to develop a new pollution standard, researched by USBM scientists and paid for by industry, was set in motion. The objective was to place this information at the disposal of the Trail smelter and the Canadians in their defense against the U.S. government!

This collusion greatly annoyed Metzger, and when the State Department threatened to invoke the Logan Act, industry backed off in July 1936. Only seven typewritten copies of a highly restricted document by Paul M. Ambrose, a chemist in the Tucson office, were produced—the one result of all this data gathering by the USBM. The overview of air pollution problems he gave provides good information on the Douglas smelter, as follows:

Sulfur recovery data supplied by Phelps Dodge and ASARCO indicated that the two border plants were emitting an estimated 1 percent of their stack gases in the form of SO_2, of which the copper plant in the El Paso operation produced 113,048 tons of SO_2 a year with none being recovered. Data furnished for Douglas show that 33,860 tons of SO_2 were recovered, but overall production data were not given. The Douglas acid plant, with its 200-ton capacity, was usually operated at one-quarter to one-third capacity. Most of the acid produced was used locally for leaching copper ores in a small, experimental operation. A small amount was being sold to other mining companies. The acid plant processed SO_2 from the roasters, the first stage in the smelting process. Dust was removed by a Cottrell baghouse system before going up the stack. The reverberatory furnaces, two of which were rebuilt in 1916 and two others in the 1930s, produced matte copper that was then processed further into blister copper in the converter plant.[24]

With recovery impractical from a business point of view, smoke damage at both border plants was settled by direct payments to claimants, the purchase of smoke easements, or the outright purchase of lands adjacent to and near the smelter. Smoke from the Douglas smelter "is usually carried long distances before it diffuses," Ambrose reported. "Fortunately, this smelter is situated in a valley that is not farmed intensively so the damage done is not of great monetary value." Unfortunately, the plume also streamed into Mexico, often at night when the winds shifted, then sometimes blew back over Agua Prieta and Douglas, its sister city. A grainy photo taken in 1936 captured a scene that would be duplicated almost every day until shutdown half a century later in 1987.[25] In short, the daily fumigations impacted American farmers growing vegetables and pecans in the Sulfur Springs Valley of Cochise County, as well as Mexican small farmers on *ejidos* (communal farms) and privately owned ranchos across the border.

Seeking to collect large awards in light of the Trail case, an enterprising Bisbee law firm represented the *ejiditarios* in their 1936 suit against Phelps Dodge for $100,000 in smoke damage. The upshot is that PD began paying annual damage awards to Mexican farmers as early as 1937. Little wonder that industry was so concerned about Trail. Here is clear proof of the first damage awards on the southern border, where the pollution problem was transnationalized by market forces decades before it became a policy issue.[26] For reasons examined in Chapter 6, the Mexican transborder linkage did not really develop until the 1980s. In fact, the Mexican payments did not become public knowledge until 1981, when in congressional testimony it was revealed that annual payments of "under $1,500 a year" were being made to farmers on both sides of the border. In 1985, Mexican farmers were said to be receiving payments of $2,007, "about the same as for the past ten years."[27]

Crop damage was being done by smelters undeniably, and the industry went to great lengths to develop legal proof that the damage was small or nonexistent. ASARCO took the lead in developing agricultural research that was used to bolster the companies' defense against damage suits in court. A standard public relations ploy was to site model farms and dairies under the lea of smelters while beautifying the plant office with a lawn and flower beds to show that it was possible to live with pollution. That most western smelters were located in remote desert areas minimized the visual wasteland of dead forests that could be seen in the early years of Trail and Anaconda. Another standard ploy was to put the onus for litigation on outsiders, the so-called "smoke lawyers," who made their living by encouraging lawsuits. But damage was being done. Recognizing that sensitive irrigated crops near its Douglas smelter were occasionally being injured by SO_2, Phelps Dodge formed in 1919 a Department of Smoke Investigation with two men trained in agriculture.

For many years, part of the mission of the PD smoke team was to advise farmers on the best varieties and growing conditions in their industrial zone. That they also gave advice on marketing conditions as a help to the farmer was often touted by the company, but this benefit cannot have been a main concern because their job was to assess the value of smoke easements, pay for crop damages, and in general handle disputes.

As field survey procedures were developed over the years, "an extensive research program has demonstrated a quantitative relationship between a reduction in functional leaf tissue caused by SO_2 and a reduction in crop yield." Plant pathologists, botanists, entomologists, range scientists, and other professionals were brought in on a consulting basis. The scientific aspects of PD's program to compensate farmers was developed primarily by Wyatt Jones and Harold Brisley. Jones's field studies on the effects of smelter smoke on plants began in California and later included work with the Selby Smelter Commission, at ASARCO with P. J. O'Gara and G. R. Hill, at Consolidated Mining Company in Canada, and with Arizona copper companies that eventually merged with Phelps Dodge. In 1929 a formal program was instituted to pay growers for crop damage, and in 1938 an experiment station was maintained at Hereford in the San Pedro Valley west of Douglas to investigate ways to substantiate economic damage. There the illusive question of invisible injury also was investigated for many years.[28]

To have any chance, however slim, of winning the legal, scientific, and public relations battles with the smelters, farmers leagued together to fight it out in state courts. Headquartered in McNeal, Arizona, the Sulfur Springs Valley Protective Association was activated in 1927 (about the same time the Northport farmers organized). Phelps Dodge responded by keeping detailed meteorological records at the Douglas plant to assist the smoke department in their work. "During the growing season (late spring, summer, and early fall) they examine all cases of damage reported and in the fall make a settlement with each farmer. They claim that the settlements are very liberal and this is probably so," a USBM staffer reported in 1934, "for since the system was inaugurated in 1930, there has been only one suit in court. The damage occurs in the Sulfur Springs Valley north of the smelter for a maximum distance of 25 miles. The company now owns or has easements on practically all tillable land within 10 miles of the smelter.[29]

In these disputes, the farmers and their attorneys typically felt outgunned. For example, in 1935 a Mrs. Frank Murphy from McNeal wrote to Senator Hayden and the USBM protesting the lack of controls on the emissions at Douglas and the impossibility of getting a fair hearing in the local court at Bisbee, which was intimidated by Phelps Dodge, or a fair settlement out of the company. Hayden referred her inquiry to the bureau for a response, which replied that dispersion was the only practical and fair solution to smoke damage. And since the USBM had no authority to investigate the problem, she was advised to seek remedy in the state courts.

Articulate and well written, her letters accurately stated the farmers' case. The Murphys had been growing vegetables for twenty years, but the damage to their crops only began in 1929, she said, when Douglas began processing low-grade copper ore. Mrs. Murphy then raised the health issue and described what it was like living with the plume: "The mountains which are ordinarily as plainly seen as if they were in your front and back yard, are entirely hidden from view and visibility for driving purposes is very poor. One gets a burning sensation in the nose and throat and it feels as if it would choke you. It has only been in recent years that the smoke has burned crops at such a great distance and with such destructive effects."

As for seeking remedy in the state courts, "you can get a hearing alright, after about three years," but this process entailed legal expenses beyond the reach of most farmers. Meanwhile, the company would settle, but for values well below the actual crop damage, on a take-it-or-leave-it basis, and only when "the farmers started a 'smoke organization.' The company then promised to pay the farmers damages and bought off some of the most active leaders by giving them smoke easements on their land for a figure which would practically cover the value of the land." If, as the bureau maintained, the disposal of sulfuric acid was not practical for western smelters, "then they should be willing to buy all the land being damaged by their smoke. They destroy grass, fences, roofs and no doubt are injuring the health of the residents of the valley." Anything as deadly to plant life would certainly be harmful to human beings, she concluded.

In closing, Mrs. Murphy asked: "How does one go about getting an investigation made of the damage done to crops and the health of the people in the affected area? The smoke damages the grass on both State and National lands—it travels for miles in every direction. It eats the galvanizing off our fences and roofs—surely it does something to the lungs and nasal passages of humans."[30] And in a follow-up letter she informed the bureau that "the Sulphur Springs Valley Protective Ass'n. intends to start a suit against the smelters this Summer and to raise the money to hire a chemist to look after the interest of the farmer and rancher but we certainly think that Uncle Sam should show some interest in the damage being done on the land owned by him."[31] In retrospect she raised the right questions, but the federal government was not about to get involved in smoke disputes on the southern border.

Agriculture and Interior both claimed they had no jurisdiction in what was a matter for the state courts to decide. "The nuisance has continued for several years," a Douglas lawyer wrote Secretary of Agriculture Henry Wallace, "and though there is much local feeling against it— the power of money in the courts, with politicians, unscrupulous henchmen, etc., makes the situation seem hopeless, so far as local effort is concerned." Since the USDA was helping Stevens County farmers in the upper Columbia Valley in their struggle with the Canadian smelter, it should now help the Cochise County farmers to fight a similar menace on the other border, he argued.

> While this smelter is in Arizona, it is affecting Mexican property across the border, and is creating a chaotic economic peril on the American side. This mining corporation has the County in its clutch—I almost said the State—and has imported hundreds of alien peon Mexican laborers to supplant American labor in the smelters— driving Americans to the land, and there destroying their crops and sustenance. We have no axe to grind about it—simply wish to see American citizens who voted for the "new deal," to get that deal.

This situation also "would require a special investigation and action from your Department."[32]

Knight, who chaired the Smelter Fumes Committee investigating smoke pollution from Trail, replied that his bureau had no direct interest in the matter. Forestry might, but otherwise "these matters are usually left to the State courts anyway and I do not believe we are in a position to do any investigating in the problem without special appropriations. The only reason we are investigating the problem in the Upper Columbia Valley is because it is an international problem and special appropriations were made to carry on the investigation." This factor was the nub of the problem, Agriculture advised. In the one case, the USDA did work under the supervision of the State Department, paid for by a special authorization of Congress, and acted only in a technical capacity rather than as an interested party. "In the case to which you refer it would seem that the trouble arises between citizens of the State of Arizona and that the proper remedy would be in the State Courts since, apparently, no Federal question is involved up to the present time."[33] And in reply to another lawyer's observation that the Sulfur Springs Valley could be developed if Phelps Dodge was obliged to put smoke eliminators on their stacks and his suggestion that Congress should compel all companies to do this, the USDA said that "there is no authority in the Department over the conditions you describe. It appears rather that any regulation or remedy of this situation must be sought from State or local authorities, inasmuch as the matter is one involving the police power of the State."[34]

Through the 1930s, followed by the wartime mobilization and the postwar boom, this situation was unchanged. Production reached an all-time record in 1942, when 1,275,322 tons of ore were smelted at the DRW. Douglas was converted to natural gas in the mid-1930s, but the change in fuel did not significantly affect emissions, and local residents seeking remedy still did so within the traditional parameters established before 1930.

Depending on the land and the distance from the smelter, easements were being purchased for at most $25 per acre. In 1971, some 111,874 acres, or around 45 percent of Sulphur Springs Valley farmland, were under smoke easements purchased by the company. "We have been buying the smoke easements up for forty years," a company lawyer said, "and . . . one of the reasons we do it is to immunize ourselves from claims for damage." In 1960, for example, Sam King Jr. of McNeal filed a typical "Easement and Release" form with Cochise County that for the payment of ten dollars released Phelps Dodge "from any past, present or future damages to the described lands and crops and to the health and comfort of the occupants thereof resulting from smelter operations, or from smoke, gasses, fumes, dusts and vapors from operation of smelters." Payments for crop damage to farms that were not under easement were made up to a maximum of thirty-five miles from the Douglas plant.[35]

For years plaintiffs lamented the expense and delays of taking on the smelters in state courts, where they seldom won. An exception was the civil suit brought in November 1954 by over two hundred plaintiffs whose lands were not covered by smoke easements. A U.S. federal court upheld PD's method of compensation. However, the farmers collected damages of $100,000 and got an arbitration procedure over future smoke damage. Although this settlement was less than 20 percent of the losses claimed for past damage to crops, starting in 1955 the smelter was placed under a regime of seasonal cutbacks during the summer growing season.

(From 1974 to 1978, sulphur emissions averaged 379 tons per day during the growing season months of July and August compared with 481 tons during the remainder of those years.[36]) Depending on humidity and wind direction and after consulting with the company's agricultural ecologist, the smelter superintendent decided when production cutbacks should occur. Crop damage in the agricultural area immediately south in Mexico was inspected "not under the decree but as a matter of good public relations."[37]

Understandably enough, those taking on the copper companies felt the system was rigged against them; the smoke department's task was to keep damage payments within bounds. The influence of Phelps Dodge at the statehouse was legendary. It was represented by the Phoenix law firm of Evans, Kitchell, and Jenkes, which also did legal work for the Arizona Mining Association. Well into the postwar years, PD controlled the *Arizona Daily Star*, which later opened its pages to a new breed of investigative reporters probing into the social and environmental problems of the industry. For their part, the farmers felt outgunned. The reason why settlements were small in the ensuing years, they believed, was because the arbiters, intimidated by the company's prestige and power, almost always accepted its assessment of damages. Bringing suit was still costly and protracted. For example, a suit brought in 1970 by the Group Against Smelter Pollution to break the legally binding power of the smoke easement system issue lingered on in federal district court for ten years before the plaintiffs gave up.

But times were changing, and the legal challenges began to include suits by urban groups worried about the health effects of smelter pollution. Tucson was known to have a much larger proportion of persons with respiratory diseases than in the United States as a whole. It was one thing to pollute remote small towns where residents traditionally regarded smoke as "the smell of money and jobs." It was quite another to impact an urban airshed already coping with automobile and particulate pollution. Beginning in 1970, the states were mandated under the Clean Air Act to set up Air Quality Management Regions. All of the smelters except two of PD's plants (Morenci and Douglas) fell within the new Phoenix-Tucson region.

It is no coincidence that the Group Against Smelter Pollution (GASP) was spearheaded originally by Tucson city councilman David Yetman, an asthmatic worried about urban air quality. Organized in four chapters, GASP was a nonprofit organization "among the purposes of which is to pursue legal means to reduce the presence of pollutants in the natural environment, and to promote the rights of all American citizens to a clean and healthful environment." The group in Green Valley represented retirees worried about air quality in the Phoenix region, while the group in

Sulphur Springs Valley was made up of farmers from Elfrida, led by Randolph "Smoky" Moses. In their class action suit in federal district court against Douglas and the easement system, Moses and other GASP members were helped by Tucson lawyers (who were also suing ASARCO and Kennecott) and a new breed of urban activists determined to "get" the smelters.

Attempting to link the health issue with economic damage, they asked, in effect, "What's coming out of the stack and is anybody being hurt?" Citing damage to his wheat and chili crops, Moses, who took on an easement when he bought his land, said, "There is smoke there most of the time at some degree or another." The smoke left a bad taste in the mouth and was "sometimes bad enough to burn your eyes . . . and brings visibility down to a quarter of a mile." Since moving to Elfrida his daughter suffered respiratory problems. But the validity of smoke easements survived this court test handily, and the first GASP eventually petered out. As a leading environmentalist recalled, their attorney was "an earth day type" who did not know environmental law, and while Yetman spouted sixties rhetoric he was not really serious. "Nothing they did since 1971 made any difference [because they were] not prepared for the hard slogging, and the tenacious, careful preparation you need to win."[38]

A decade later Phelps Dodge said it was paying about $5,000 a year for crop damage to farmers both in the United States and Mexico. A company official was quoted as saying that the very small sums involved were a sign of how minimal the damage being caused by smelter smoke really was. However, starting in 1981 and amidst the rising public criticism of Phelps Dodge over smelter smoke, the farmers found a more favorable climate for their suits. In 1982 the Mitchell family in Elfrida sued PD for having damaged land and pecan trees with "unreasonably high and dangerous levels of pollution" between 1979 and 1981. At least some of the Mitchells had signed a smoke easement. Even so, Mitchell Pecan eventually won a $150,000 settlement against Phelps Dodge.[39]

Meanwhile, new research into the effect of SO_2 damage on plants had raised doubts about the industry's assumptions on the safe threshold for the amount and frequency of fumigations established by Hill and Thomas at the ASARCO research station in the 1930s. As confidently affirmed in the ASARCO company history, from this research "emerged a fundamental fact. SO_2 was injurious only when it affected or touched plants in a concentration of over 3 parts per 1,000,000 parts for a continuation of three hours in daylight and in humidity exceeding 50 degrees. It had been discovered that at night it took nearly double the concentration to do injury to vegetation." ASARCO was still affirming these results, as bolstered by Thomas's subsequent research on the effects of fumigations on vegetation published in a paper circulated by the industry in 1971. Their scientist concluded that the federal emissions standards were "far more than adequate to protect against injury to vegetation."[40]

But in the 1970s the invisible injury thesis resurfaced, with its emphasis on cumulative injury from exposures to even low levels of SO_2. (Recall how assiduously, and with what success, the Canadian lawyers and scientists, aided by the American smelter industry, had striven to discredit this thesis.) The Arizona press were briefed by Wayne Williams, a plant pathologist, who had been retained by a local environmental group (headed by Dick Kamp) to assess crop damage on the Mexican side of the line. Williams cited the myths under which big industry operates when they talk about crop damage. The first myth "was that the only damage to a plant is what you can visually verify, such as injured leaves. Inspectors don't acknowledge the fact that we now know that sulfur dioxide can retard growth 25 to 30 percent in plants without showing any visible signs of injury." And the second myth was that there is no SO_2 damage to the pores of plants at night because their pores are closed. From research done by British scientists and published in the journal *Environmental Pollution*, "we now know that plants do get injured at night. Small amounts of sulfur dioxide can actually cause the pores of plants to open at night, thus leaving them more susceptible to injury," Williams said.[41]

As mentioned, Phelps Dodge maintained a small research station on the banks of the San Pedro River near Hereford until the early 1980s where controlled field experiments were conducted to study the effects of sulfur dioxide on crops. This station contributed to the old debate about local damage to crops around smelters. "Subtle effects of SO_2 (invisible injury) have been the focus of numerous experimental fumigations . . . [company scientists maintained], but no crop yield loss has been attributed to the effects of SO_2 unless visible foliar injury was also observed." If this was so, why was research into the low-level effects continued for so long at Hereford? They even cited a 1961 paper by M. D. Thomas, Hill's old mentor, that an experimental procedure might be devised that would produce significant injury from SO_2 without visible effects, "but that it would probably be of academic interest only."[42]

However, new data from the 1970s were being used to rehabilitate the invisible injury thesis, even if it was still difficult to use in sustaining proof in damage awards. A University of Arizona researcher testified in the above-mentioned GASP smelter case that the only generally agreed upon method to assess leaf damage was to calibrate the amount of visible injury.[43] What was really new was the rapidly growing alarm over the long-range transport of sulfates, known as acid rain, to which this smelter was a large contributor.

Although much was still unknown, research over the last decade had shown that yield losses could occur in the absence of foliar injury after long-term exposure at low concentrations. If the scientific community was cautious—a 1984 conference on acid rain held at the Royal Society in London concluded that "we are far short of a basis for estimating subtle damage below the threshold of visible injury in the field"—the public and the politicians were primed to hear more bad news when the effects of acid rain began to be observed by scientists and commented on in the popular press.[44]

Recent research on the effects of SO_2 from midwestern power plants was showing that even short-term fumigations of two parts per million could cause damage. Private laboratories, the EPA, and the Environmental Defense Fund's scientists were all publishing results that did not support the industry position of a higher pollution standard. Given the rising public alarm over acid rain, it may well be that winning damage suits would become less likely for a company like Phelps Dodge. However, the issue soon became moot after the shutdown of the three uncontrolled copper smelters that could still cause this sort of damage: ASARCO's Tacoma, Washington, smelter, Kennecott's McGill, Utah, smelter, and the Douglas Reduction Works.

Society was changing rapidly, and well-educated constituents imbued with an ecological perspective were becoming active in a changed legal climate. Broad new standards for protecting public health and welfare now took precedence over the older emphasis on establishing proof of damage in air pollution cases, thanks to the passage of the Clean Air Act as amended in 1970. Smelters that could not meet the new performance standards for SO_2 and particulate emissions would have to modernize or shut down.

By 1970 it was clear that the federal government would be cracking down on sulfur dioxide emissions by copper smelters. The companies had two choices: the capital equipment route, which meant installing expensive abatement technology, or the litigation route, which meant stalling for time to avoid or delay shutdowns for noncompliance while the new clean air standards were being tested in court. Phelps Dodge elected a combination of both. It built a modern smelter using Finnish flash-furnace technology at Playas, New Mexico, while continuing to operate three old smelters at Morenci, Ajo, and Douglas employing conventional reverberatory furnaces. Of these, Douglas was by far the oldest, having been upgraded in 1916 and again in the 1930s. Efforts to modernize the Morenci and Ajo smelters were abandoned, and after years of litigation these plants were eventually shut down, Morenci and Ajo in 1985 and Douglas in 1987.[45] The strategy for Douglas was to keep this low-cost producer operating as long as possible under a provision of the Clean Air Act that provided a cumbersome mechanism called a nonferrous smelting order (NSO) for short-term exemptions on economic grounds (see below). A joke about the drawn-out effort to define the terms of Douglas's compliance with the national air quality standards made the rounds:

Question: What *is* a variance?
Answer: A good lawyer and a tall stack.

To bring the three PD smelters into compliance with the new air standards, acid plants were installed in the mid-1970s in the first two along with the other large smelters in Arizona in order to meet the act's requirement for constant controls. However, plans to install a sulfur recovery system at Douglas were abandoned. Intermittent control systems (SCS) were also installed at all these plants, including Douglas on August 31, 1975. Much less costly to operate, along with tall stacks this older dispersion system was permitted under the act as a temporary measure in conjunction with constant controls such as the acid plants.

Because of the high costs of modernizing Douglas—estimates ranged between $250 million and $500 million—Phelps Dodge preferred the status quo. However, in May 1970 the company did assure the Arizona State Board of Health (which was the agency in charge of setting emission standards and enforcement) that it would reduce input tonnages by about 50 percent by cutting the reverb furnaces back to two and eliminating roasting, which produced about half of the SO_2. PD also agreed to install electrostatic precipitators to cut down on particulates and scrubbers to clean off-gases from the reverbs and converters. Given the poor market for sulfuric acid, Phelps Dodge and ASARCO built an experimental $2 million pilot plant at El Paso designed to produce elemental sulfur, which was much easier to handle and ship. Evidently this pilot did not pan out, and the DMA acid absorption plant was not installed at Douglas.[46]

The basic technical problem then and later was that the gas stream from the smelter reverberatory furnaces used at Douglas was not rich enough in SO_2 for an acid plant to operate efficiently. At less than 4 percent concentration, the smelter off-gases were simply too weak for use in conventional acid plants. Trail solved this concentration problem by installing new technologies at its zinc and lead smelters in the 1920s. Modernizing Douglas was not part of the business plan to save Phelps Dodge in the 1980s.

Uncontrolled, Douglas was a low-cost producer well designed to smelt a variety of concentrates. In contrast, the new Hidalgo smelter using flash-furnace technology was controlled, cost more to operate, and was subject to shutdowns and exceedences until the system was fully worked out.[47] The problems at Hidalgo were one major reason why Phelps Dodge fought so assiduously and for so long to keep Douglas operating.

In 1971 PD petitioned to achieve compliance under the new federal standards, which it claimed were both more flexible and less stringent than the state standards mandating "a rigid 90 percent removal requirement." Complying with the state standard would require new capital costs of $249 million. However, by using the federal approach that permitted achieving emissions limitations with less restrictive controls, coupled with restricted operations, the new federal air standards could be achieved. The company planned to achieve these results by using the closed-loop emission limitation system (SCS), a combination of reductions in the amount of sulfur in the smelter feed and periodic curtailment of operations at Douglas depending on meteorological conditions. By 1973 Phelps Dodge was well down this road, with plans to spend a total of $17 million on its air quality control program, which included a permanent partial curtailment of operations under an operating license that the state extended until the end of November 1974.[48]

By using only SCS, Douglas operated without constant controls such as an acid plant and scrubbers, until closure at the end of 1986. From the early 1970s on, Phelps Dodge always said that modernizing Douglas was not an option; instead, if forced to, it would shut down, leaving Douglas without some four hundred jobs and a $10 million payroll. Singled out as the nation's worst polluting smelter, in 1974 Douglas accounted for 25 percent of the entire state's daily emissions of SO$_2$—1,249.6 out of 4,810.4 tons per day. Virtually no sulfur was recovered from the Douglas operation, and emissions only went down in the early 1980s because of temporary shutdowns due to the strike and a weakening market for copper.[49] With price recovery, emissions went up again.

Why was the antiquated Douglas smelter, last modernized in the 1930s, kept open and Morenci, built during the Second World War and serving the large, expanded Metcalf mine, shut down? In part the answer is that DRW, the largest smelter in the PD system, was used to process ores from different sources, most recently from Bisbee until the Lavender Pit closed in the mid-1970s. Afterward it was used as a custom smelter—meaning Phelps Dodge owned the concentrates and the products at the other end—and to process overflow from PD mines.[50] It also served as a toll smelter, processing the ores from other companies. Retaining this flexibility was useful while the company repositioned itself to deal with the new and costly environmental regulations and coped with the decline in copper prices beginning after the oil shocks of the 1970s, which also greatly increased the price of natural gas used in the furnaces. In part, also, it was because Phelps Dodge in the early 1980s was confronting a major strike and a serious cash flow problem. Struggling to survive, the company simply ran the antiquated Douglas works as long as possible.

Table 2. Douglas Sulfur Emissions

	Tons/Day	% Sulfur Removal	All AZ Smelters, Tons/Day	% Sulfur Removal
1972	686	2.5	2787	15.7
1973	721	2.3	2890	15.5
1974	625	2.3	2385	20.8
1975	663	2.2	1818	37.3
1976	534	1.0	1382	49.5
1977	275	2.7	1009	55.0
1978	249	2.4	1061	54.9
1979	361	1.9	1209	55.0
1980	330	1.8	874	50.6
1981	455	1.6	1197	53.5
1982	124	1.4	563	58.6
1983	254	1.7	672	50.6
1984	438	2.9	833	58.8
1985	421	1.9	753	59.9
1986	339	1.5	650	64.0
1987	12	2.0	303	79.3
1988	0	0	176	88.0

Note: One ton of sulfur equals two tons of SO_2.
Source: Table, "Trends in Arizona Copper Smelter Sulfur Emissions, January 26, 1989," in Larry Bowerman, "Copper Smelting in EPA, Region 9," long typewritten synopsis, in file Bowerman A-53, EPA Region 9.

Also on the technical horizon was the possibility that leach mining, an updated version of an old process once tried experimentally at Douglas and once used at Ajo and Pilares de Nacozari, could prove to be a cost-effective way to process the mountains of calcite ore that had accumulated at Morenci and Bisbee during open pit operations. The smelting of low-grade sulfide ores in the reverberatory furnaces at Douglas would continue as long as possible, now that smelters at Morenci and Ajo were closed, and then be processed at Hidalgo in New Mexico, which is what happened in the late 1980s. However energy-inefficient, the DRW plant was cheaper to run than building a new controlled facility. Even though Douglas with its antiquated equipment was not fuel-efficient, the plant was long since amortized, and it made sense to keep it open until leaching came on line.

The big uncertainties in bringing the PD plants into compliance with the Clean Air Act were technology and costs. To be sure, this is a conservative industry, slow to change by nature of the fact that proving up and developing ore bodies is a long-term proposition. Moreover, having controlled the state for so many years, when faced by the new federal regulatory climate management was also by nature cautious and reactive. Opponents saw this response as playing for time (which in fact it was) until the new laws could be weakened or scrapped. That new furnace technologies were cost-effective and available—the Japanese had installed them after being told by the government that abatement took precedence over considerations of cost in marketing the by-products, that they had to clean up[51]—was well established. Once under the regulatory gun, Phelps Dodge had to respond.

The business solution was to build a new plant, which would process ore from Tyrone, at Hidalgo in the Playas Valley on the New Mexico border while attempting to modernize the old "reverbs" installed at Morenci during the war and the Ajo smelter. Budgeted initially to cost in excess of $100 million, the Hidalgo complex with its smelter using a modern Outokumpu flash furnace of Finnish design eventually cost $268 million, which included laying out the town, the railroad, and the concentrator. Of this total, an estimated $100 million was for pollution control. When the new facility came on line in 1976, it was also receiving ore from the Cyprus Baghdad copper mine in Arizona, which also invested in the plant.[52]

Using an oxygen injection system, the flash furnace could smelt lower grade concentrates into matte by using the concentrate's own content of sulfur to create the heat needed for smelting. Converters then transformed this matte into blister copper that, after further processing into anode copper, was shipped to El Paso to be refined. Roasting was eliminated altogether. Flash-furnace technology could also produce blister directly from concentrates, using a closed system less susceptible to fugitive emissions than the Playas smelter. A drawback of the Outokumpu process was the added expense of treating slag. Used in conjunction with primary and secondary acid plants, well over 90 percent of the SO_2 could be recovered. As it happened, Phelps Dodge had trouble operating this plant, and gases were periodically vented during repairs until the state of New Mexico cracked down.[53] In 1987, the company purchased Kennecott's new flash-furnace smelter at Chino, New Mexico, which processed higher grade ores. PD ended the decade with two modern smelters in New Mexico, no smelters in Arizona, and with large leach operations in place.

By the early 1980s, ASARCO had modernized its Hayden plant with the Canadian-developed INCO process, which it later used in cleaning up the El Paso plant as well. Kennecott and Inspiration, a Canadian-owned company, both had modern plants. But attempts to modernize Morenci and Ajo did not work out (see below). By the early 1980s, flash-furnace technology had been successfully applied to convert traditional reverberatory furnaces at CODELCO's smelter in Chile and later at the Magma Copper Company's San Manuel smelter in Arizona. Magma reportedly spent over $200 million to modernize the San Manuel complex, of which some $130 million was devoted to the smelter itself.[54]

Morenci smelter, which served the expanded Metcalf mine, PD's jewel in the crown, was modified between 1967 and 1975 in response to environmental legislation. Converter gas was captured in electrostatic precipitators and waste heat coolers, and a single contact sulfuric acid plant began treating all converter gas in 1974. As well, the Morenci smelter began operating under a supplementary control system on January 1, 1976. The four "reverbs" constructed at Morenci during the war and a fifth added in 1974 were fired with natural gas. Reverberatory off-gas was electrostatically cleaned and then vented through a 603-foot stack. In and of themselves, these improvements were not enough to meet mandated standards by the state, and in 1981 the company signed a consent decree with the EPA that stipulated a plan to achieve particulate and SO_2 reduction targets by January 1, 1985. PD was also obligated to commit an estimated $150 million to the Morenci smelter project.[55]

The plan included the application of oxygen enrichment on two furnaces and improvements to the acid plant and other gas systems. Three of the old wartime furnaces were to be discontinued except for low-sulfur ores. In the interim, Morenci was to achieve at least 62 percent overall sulfur removal in the years 1982, 1983, and 1984, which meant that in order to meet these targets Morenci had to *reduce* throughput just as new capacity from the Metcalf mine and Tyrone came on line.

Aware of CODELCO-Chile's successful application of oxygen-fuel burners to a reverberatory furnace at the Caletones smelter, Phelps Dodge had begun oxygen-fuel burner testing at Morenci during the summer of 1979. Initial tests were promising. "The oxygen-fuel process offered an immediate opportunity to reduce energy consumption while increasing furnace gas sulfur dioxide concentration sufficient to accommodate capture in the acid plant."[56] From the enforcement point of view, things were going well: the abatement order seemed "to be having the desired effect, since there has been only one NAAQS [National Ambient Air Quality Standard] violation since the order was issued," and conversion coupled with other changes was proceeding, an EPA official observed.[57]

However, the system did not pan out, and PD elected not to press on with the modernization of Morenci. After expenditures of $89 million, in February 1985 the company decided that the oxygen-sprinkle system was unsuitable for Morenci, and the smelter was shut down permanently. Only one of the modified furnaces had been brought into commercial operation; construction to install oxygen-fuel and oxygen-concentrate burners on the number five furnace was stopped in October 1983, and it was deactivated one year later. Ajo was also shut down.

It is important to underscore once again that the Douglas smelter did not have a place in this long-term business strategy. It had been built decades ago to service mines in Mexico and Bisbee that were now no longer big producers for Phelps Dodge. Given transportation costs, locating a smelter at Douglas was no longer optimal from the standpoint of assembling raw materials; building a new smelter at Douglas would not make economic sense. But the old smelter, decapitalized and patched together as it was in the later years, had a useful place in the transition to clean smelting and leach mining.[58] At a time of financial crisis for the corporation, Douglas did indeed live up to its nickname of "Old Reliable"—from the business point of view.

Facing a severe cash flow problem in the early 1980s, Phelps Dodge stopped spending funds to modernize the smelter at Morenci and switched instead to the newly developed surface leaching system (SX-EW) that could process calcite ores cheaply using large amounts of sulfuric acid. Leaching was also an answer to the acid disposal problem that had long plagued the smelters. By avoiding crushing, smelting, and refining, the leach process could produce a pound of copper from waste rock for less than thirty cents. As well, PD borrowed heavily to modernize and expand capacity at the Morenci-Metcalf mining complex and at Tyrone. At this time, Sumitomo took a 15 percent share of the Metcalf mine, which brought in the cash to fund the new leaching operation. Sumitomo also smelted some of the Metcalf concentrates, the bulk of which were sent to Douglas. British Petroleum, the new owners of Kennecott Copper, sold PD the modern Chino smelter and mine, in which Mitsubishi already had a third interest. PD also won the big strike at Morenci, and by the time copper prices rose in 1987 its Arizona labor force was entirely nonunion. Having embarked on a drastic restructuring, the core of that survival strategy was "closing old, inefficient smelters, looking for partners to inject new cash, investing in new technology—and sheer ruthlessness." Thanks to this successful business strategy, Phelps Dodge emerged from the 1981–1986 depression years for copper as the nation's largest copper company.[59]

From the company point of view, "the impact of the costs for pollution control came at the worst possible time for Phelps Dodge. With copper prices plunging in 1981–1984, the company's losses were staggering." The drain on capital because of pollution controls complicated an already difficult business problem. Cash dividends on common stock that had been paid every year since 1933 were suspended, and by December 31, 1984, the once debt-free company's total long-term debt was $591 million, of which $216 million was air quality control obligations. Thus, "the impact, primarily on the smelting phase of production, was even greater than it might otherwise have been because the standards were imposed primarily during a time when world copper prices were already falling and U.S. [copper] firms were under heavy pressure from other rising costs." Nevertheless, as Hildebrand and Magnum conclude in their sympathetic survey of the industry, "By closing outmoded smelters, remodeling where possible, and replacing where not, copper smelters reduced their sulfur dioxide emissions by three-fourths, achieving a capture rate of 90 percent."[60]

From the public policy point of view, absent a stringent regulatory climate, smelters that had been venting SO_2 into the air for decades as the method of choice for disposing of unwanted by-products had no incentive to change. A case in point is the Southern Peru Copper Company's smelter at Ilo, Peru, in which PD has a minority interest with ASARCO. Until recently, they had run this smelter without controls. In 1995, this smelter reportedly was emitting two thousand tons of sulfur dioxide a day, "causing smoke so thick it hovers over the city like heavy fog and sends residents to hospitals, coughing, wheezing and vomiting."[61] Industrially proven flash-furnace technology had been around for decades, but the U.S. smelters were slow to adopt it. Phelps Dodge only did so at Hidalgo after passage of the Clean Air Act.

Faced with a drastically changed regulatory climate in the early 1970s, the initial response of western copper companies was to employ halfway measures such as SCS and tacked-on acid plants to control some of the pollution while tying up regulators in the courts and fighting them through allies in Congress. A confrontational approach was the tactic of choice, in part because the costs of compliance were so high.

A combination of the new enforcement regime, grassroots activism, changing demographics, and the new science backstopping the Clean Air Act contributed to bringing the smelters into compliance, which is the subject of my next chapter. Suffice it to say here that their physical and social isolation made the smelters a target for environmental activists and regulators more readily than, say, the electric power industry or for that matter the petroleum-based automobile culture that still holds America in thrall. In practical terms, it was easier to control single-source polluters like copper smelters. The diffused airshed of power plants was less suited to a command and control approach; pollution caps, administered through the innovative mechanism of freely traded pollution permits, have proved effective in recent years.

For illustrating both the business and the public policy perspectives, the case of the Douglas smelter is made to order. For as long as possible, Phelps Dodge did everything to maintain production at its most antiquated smelter, the facility that the company called "Old Reliable," but others referred to as "Old Smokey," the single most polluting industrial plant in all of North America. Operating at full capacity, it could produce up to 180,000 to 200,000 tons of anode copper annually. An illustration of just how valuable Douglas was to the system is the comparison with Hidalgo, which could produce 225,000 tons of anode copper. Having been a pillar of the PD system since 1900, Douglas was a workhorse that smelted concentrate from Bisbee, Nacozari, Ajo, and Jerome. Until the 1970s it was slated to handle ore from Tyrone until passage of the Clean Air Act prompted Phelps Dodge to construct the flash furnace at Hidalgo instead.

By 1981 it was clear that Douglas "was under environmental attack and could be forced to close by the end of 1987 for not meeting clean air rules."[62] However, by then it was picking up some of the overflow from Morenci when that smelter was first curtailed and then shut down. Another way of answering why Phelps Dodge fought to keep Douglas open must be that it paid to pollute—over the smelter's long life and perhaps particularly during the uncertain restructuring years of the 1980s. But bear in mind that there was not a good short-term alternative to keeping Douglas open.

Phelps Dodge said all along that it was not worth the estimated $500 million cost to retrofit Douglas, and building a new plant in that location was not viable. Under pressure to meet national ambient air quality standards (NAAQS) as mandated by Arizona's state implementation plan (SIP), Douglas was the first Arizona smelter to operate a Supplemental Control System (SCS), beginning in August 1975 with Morenci and Ajo the next January. The primary purpose of SCS at Douglas was "to maintain Arizona Ambient Air Quality SO_2 standards by curtailing emissions from the smelter during periods when meteorological conditions exist or are forecast to occur, which could cause these standards to be exceeded." The system went through three iterations, having first been modified to comply with the hourly averages imposed in 1982, and then was to be revised again in 1986 to better capture fugitive emissions and to deal with short-term peak exceedences. But what had been an innovative response at Trail in the late 1930s did not measure up to the EPA's much more exacting standards fifty years later. The revision was rejected.[63] To attain compliance before shutting down in 1988 when its operating permit was due to expire, PD would have had to curtail production even further. In July 1986 it elected to shut down.

The copper companies argued that SCS would enable them to meet the air quality standards mandated by the state's SIP. This SIP was less stringent than EPA guidelines required and more favorable to the companies than the EPA would have liked. Achieving compliance in a state long beholden to the copper companies was a tough battle for the EPA, as covered in the next chapter. The EPA also took a hard line on control technology, arguing successfully in federal court that SCS could only be used on a temporary, transitional basis "after the installation of reasonably available control technology with the companion requirement that the ambient air quality standards ultimately are to be achieved through the use of positive control alone."[64] Stack emissions at Douglas were reduced by half, almost entirely by means of reducing throughput. Yet in 1984 the EPA considered that "the uncontrolled Phelps Dodge smelter" still accounted for 53 percent of total Arizona smelter sulfur dioxide emissions.[65]

The copper companies and their allies in Congress responded by claiming that the industry was severely depressed "due to excessive production by government-owned foreign competition," namely by the recently nationalized Chilean mines, and needed more time to achieve compliance with emission limitations on SO_2 and particulates. Section 119 of the 1977 amended Clean Air Act allowed smelters to delay compliance by using intermittent control systems to curtail production *but only as a supplement to continuous emission controls*, such as sulfuric acid plants. Under section 119, companies could apply for a "nonferrous smelter order" (NSO), renewable once and the second to expire on January 1, 1988. As well, Douglas and Kennecott's McGill, Nevada, smelter were further exempted, being permitted to use intermittent controls alone to meet the Clean Air Act standards upon a showing that the requirement to install continuous emission controls "*would be so costly as to necessitate permanent or prolonged cessation of operations of the smelter.*"[66] It was the granting for Douglas of the first NSO by the state in 1982, followed by PD's request for renewal, that stirred a hornet's nest of opposition.

The EPA opposed the NSO approach all along and battled through the courts for consent decrees to impose implementation programs for permanent controls on each smelter not in compliance, requiring tight schedules with deadlines and imposing steep fines for nonattainment. PD battled for a second NSO, which would have extended the life of Douglas until 1988, but at the cost of having to post a $3 million bond as an assurance of compliance and of accepting a fine of $400,000 for past air quality violations. Having concluded that the new attainment targets would necessitate further production cutbacks—another 10 percent with the smelter already operating at only about 60 percent of capacity—a company official informed the EPA that "Phelps Dodge's rational choice would be to close the Douglas Reduction Works smelter rather than attempt to operate it on a one-furnace basis." Operating with one furnace, smelting costs at Douglas would be $106 per ton for Morenci concentrate, while another domestic smelter with costs of $84 in combination with a foreign smelter costing $86 per ton would be less expensive. Even with much lower freight rates for shipping concentrates because of its location, a pared-down Douglas with operating costs of $96 per ton became a relatively high-cost operation.[67]

Delaying the Douglas shutdown had worked for PD: it bought more time for the Morenci SX-EW leaching operation and processing plant to come on line. Leach operations were also installed at Chino and Tyrone. In the meantime, two ASARCO smelters could process the concentrate from Morenci that had formerly been smelted by DRW.[68] From the company's perspective, hardball tactics had won the results expected of them, short of continuing operations indefinitely at DRW.

From the public policy perspective, the EPA and the environmentalists had won the war to control air pollution from smelter smoke. In truth, their victory cannot be disentangled from the business strategy adopted to solve the serious problems faced by Phelps Dodge. Thus the other side of the story is that Phelps Dodge emerged from this decade of strife as a much more productive and competitive operation. Does this particular outcome bear out an environmental maxim: that good environmental management leads to better cost controls, better use of technology, and more competitiveness? My answer is a qualified yes. Phelps Dodge emerged from the twin challenges of depressed copper prices and pollution abatement a stronger company. But keeping Douglas open had been part of that strategy. The workforce at this old plant knew how to run it supremely well. There was no indication that it would be shut down voluntarily anytime soon.

PD's turnaround won accolades in the business community but few friends among the public at large, which was reeling from the effects of employment downsizing and union busting, of which this company was a prime example. Furthermore, PD's confrontational tactics were no longer supported by ASARCO and Magma, which by the early 1980s were taking a conciliatory policy toward the EPA and the more moderate NGOs.[69] For their part, environmentalists and regulators joined battle with what they thought was an industry in decline, which in any event was not so. However, from the activists' perspective, tagging a "heavy" like Phelps Dodge with the perception of being a loser and seeing it as a reactive giant that could be bested by using new tactics worked well, as shall be discussed next.

As for Douglas, the loss of 390 jobs, while serious for the workers, was soon absorbed by the town with its new prison and changing economic base. Total shutdown costs of $10 million were written off, and a plant closure plan was initiated in the fourth quarter of 1986. Total write-off accrued in 1986 was $8,245,000, of which the largest single item was the pension plan, followed by the net book value of $2.7 million and surplus equipment valued at $2.6 million. Equipment including three converters was sold to Cia. Minera de Cananea for $450,000. Among the other obligations in the shutdown costs were $150,000 for litigation costs and settlement of the Mitchell Pecan lawsuit, the largest single payment in the history of Douglas. Another substantial charge was $600,000 for environmental penalties.[70]

Reflecting sentiments still very much alive in the old mining center the city of Bisbee declared a one-hour memorial on January 15, 1987, to commemorate the closing of the Douglas Reduction Works: "This occasion is intended to reflect upon the many benefits bestowed upon Bisbee by the Phelps-Dodge Corporation. Their contributions have touched the lives of every resident, present and past. And so it is, fitting, that at the end of an era, we collectively give thanks for all they have done."[71]

It was the passing of an era. When I informed old family friends from Dawson and Clifton about the subject of this book they responded, "Don't be too hard on Phelps Dodge." This is the message I hear from my grandfather's day. Turning next to the regulators and their sometime allies, the environmentalists, it bears repeating that the business side of this story matters a good deal too.

CHAPTER FIVE

Mining Is Rarely a Local Event

The Clean Air Act as amended in 1970 set very ambitious federal standards for controlling air pollution, leaving implementation to the states. The two basic mechanisms of this regulatory program were the establishment of the national ambient air quality standards (NAAQS) for certain emissions—in the case of smelters, SO_2 and particulates—and the state implementation plan (SIP), which was subject to EPA approval. Several sources generating airborne pollutants were targeted, among them copper smelters, of which the Douglas Reduction Works was one of the most challenging to the regulators seeking to bring these facilities into compliance with the Clean Air Act.

On the federal side, enforceable requirements were established based on three objectives: to achieve a level of ambient air quality necessary to protect public health, to achieve within a "reasonable" time a level of ambient air quality to protect public welfare (defined in the act as "any effects on soils, waters, crops, vegetation, man made materials, animal welfare, weather, visibility and climate, damage to or deterioration of property and hazards to transportation, as well as effects on economic values and on personal comfort and well being"), and to maintain existing ambient air quality in air that was nonpolluted. Thus there were three indices for ambient air: the primary standard, providing for the attainment by 1975 of an air quality level that was not detrimental to public health; the secondary standard, calling for air of a quality so that there would be no detrimental effects on public welfare from air pollution within a reasonable period after 1975; and the nondeterioration standard, which guaranteed no significant downgrading of air that already exceeded primary and secondary standards.

The state plans were responsible for such items as individual source emission limits, schedules and timetables for compliance, control technologies to be used in meeting these limits, and methods of monitoring and enforcement. Each SIP had to be submitted for approval by the EPA and was subject to revision as the rapidly evolving scientific and engineering knowledge of air pollution was developed.[1]

With respect to smelters, the primary objective of the act was to control ground-level emissions of SO_2 and particulates at the source. Given the isolated location of these plants, it was not difficult to identify point-source pollution and thus assign responsibility for damage. The smelters had accepted claims for damage to property all along, and since the Selby decision in 1914 the courts had established emissions limits on certain smelters, in Utah, Montana, and (after international arbitration) at Trail. In 1955 a regime was mandated for Douglas during the growing season. For decades the smelters had been willing to pay for the necessity—from their perspective—or the privilege—from the perspective of those who believed air was not a free public good—to pollute. Now the rules of the game had changed radically. "The pragmatic, functional definition of air quality, restricted to what was economically and technologically feasible, was abandoned, and clean air was legislated a fundamental national value."[2]

After 1970 it was necessary to comply with new, national standards that were designed to protect health, with particular attention to vulnerable individuals such as bronchial asthmatics and emphysematics, old people, and children, and the general welfare, which included protecting the basic integrity of the biosphere. To the traditional economic considerations, which were limited to the actual damage to property and took into account economic factors (profitability, availability of technology) in determining the level of control applied, were added health and welfare criteria that reflected both what had recently been learned on the science side about the effects of air pollution and public concerns about the deteriorating quality of life, especially in urban areas.

Furthermore, in response to public anxiety over air and water pollution, federal jurisdiction was strengthened as part of a general movement to extend the national government's regulatory powers. Created in 1971 by the Nixon administration, the Environmental Protection Agency was given broad authority to set standards and enforce them. It was not a question of sharing responsibility for enforcement with the states; this was not a federal-state partnership. In fact, "the granting of broad enforcement authority to the federal government was a significant shift of responsibility from the state to the federal level."[3] With respect to the smelters, this change effectively got the states to react, because up until then the governors were in the hands of the smelter interests. "With passage of the Clean Air Act, they lost their edge, and we wore them down," the EPA's David Howekamp told the author.[4]

The EPA administrator had broad authority to disapprove state plans that did not meet federal criteria, which he proceeded to do in a series of challenges to Arizona's SIP, the first element of the enforcement story that will be examined here. One is struck, in reading through the records of this drawn-out struggle, just how public and transparent it was. This was the style and spirit that Senator Edwin P. Muskie and others had built into the act. It is well to keep this point in mind in looking first at the governmental side before turning later to the role of the environmentalists.

Among the government players, certain individuals stand out. At the EPA, David P. Howekamp, the Air Management Division head of Region 9 in San Francisco, was prepared for a long campaign that he waged relentlessly with the aid of legal officer John D. Rothman and technical expert David A. Solomon. In Washington, Rachel M. Hopp did the legal work at the Air Enforcement Division. Sitting next to Howekamp in hearing after hearing, Solomon, with his deep-set brown eyes and flowing black beard, seemed to emanate a sort of biblical presence proclaiming "Thou shalt not pollute." Other important players on the government side included Ed Reisch in the national enforcement program, who was deeply committed to the smelters issue. Larry Bowerman worked for thirteen years on smelter issues at the Air Management Division in Region 9. Lee Lockie, director of environmental quality at the Arizona Department of Health Services (ADHS), was an expert on the smelter monitoring issue. At first a reluctant partner with the federal air enforcement division, ADHS came to see eye to eye with the EPA.

The Phelps Dodge officers most responsible for dealing with them included Senior Vice President Leonard Judd and Matthew P. "Pat" Scanlon, another vice president who was Howekamp's counterpart. Scanlon grew up in Dawson, where his parents worked in the Phelps Dodge Mercantile, and was the first Dawsonite to attend Harvard University. He was assistant to the president in Phoenix during most of the 1980s. Attorneys from the Tucson firm of Evans, Kitchel, and Jenckes and from the Washington firm of Verner, Lipfert, Bernhard, McPherson, and Hand did the legal work. James Bush and his assistant, Amy R. Coy, appeared often for the company in cases being heard by the Ninth U.S. District Court in Tucson, which had jurisdiction. Jack Boland, another PD lawyer, was a specialist on section 119 of the Clean Air Act.

From PD's point of view, the company's survival was at stake—because of the strike, the fall in copper prices, rising debt, and the capital costs of environmental controls. "I have represented Phelps Dodge in Arizona for 25 years," Bush wrote a congressman in 1986, "and the last three have been by far the most difficult." From the governor's office, a former staffer recalled, "Bush's job was to play for time. But they never acted as though they would close the smelter," which of course they eventually did. From the EPA an air pollution expert reflected on these years: "A real challenge of the job was dealing with some of their best technical and legal people. It was scary for a young guy. But I felt if I did my homework and had my facts straight I could hold my own in meetings with these folks. We persisted. With good data we could take them on."[5]

As soon became apparent, the act was long on goal setting but weak on the specifics of implementation. The result was that much of the progress in cleaning up the air was established by court test, the result of seemingly endless litigation. As Karen Ingram points out, the setting of high goals without giving much attention to efficient implementation was no aberration in the political climate of the times. In her words, "innovative legislation—setting targets considerably more ambitious than what was known to be achievable—made sense."[6]

Given this approach, implementation of the Clean Air Act was slower than expected, usually adversarial, and sometimes contradictory. It was deliberately pragmatic in the latitude given polluters to achieve the technology-forcing goals built into the act and allowed for a certain flexibility in responding to what was concurrently being learned about the physical effects of air pollution as well as what was technologically feasible to abate it. As Muskie pointed out, the choice of technology was left to the companies: "These laws demand only that the pollution controls be enforceable on a continuous basis against precisely defined criteria." But because acceptance was slower than expected, "thus far, our reliance on performance standards has been only partially adequate. . . . We have come only a small part of the way in developing an environmental ethic, toward the best and the cheapest ways to transform pollution to a recovered resource rather than a discharged waste."[7]

The traditional economic performance standard, based on what was both affordable and technically feasible, was the pathway preferred by business. Initially, there was strong resistance to the new environmental laws. The initial tactic of the smelters was to tie up in court proceedings the enforcement of laws that they felt would not stick and would then go away. But by the mid-1970s they knew the laws would stick, and they would have to deal. "Most chose the 'wait and see' approach and decided to accept temporary reduction of their production [using SCS] in order to minimize or postpone completely the major expense of installing pollution control equipment," a mining executive wrote. (Plants like Inspiration's Miami smelter that did install constant controls were put at a competitive disadvantage.[8]) Mostly they evaded the issue of permanent SO_2 controls, and it would be a decade after passage of the 1970 version of the act before Arizona had its final SO_2 implementation regulations in place. Far from developing an environmental ethic, the mining interests, when faced with the challenge to upgrade, saw huge capital costs ahead and unwanted sulfur by-products on their hands. They did build new facilities, including PD's Hidalgo smelter. But the old smelters were already paid for and were profitable, the Douglas Reduction Works (DRW) being the leading example.

Another feature of the act was the wide scope given to public comment and participation. The upshot was that previously isolated local groups could seek each other out at public hearings, in lawsuits, in generating information—all under the umbrella of the act. Local groups welcomed the opportunity to participate in EPA hearings. "We're glad you're here," Michael Gregory, a Bisbee environmentalist, said at a 1979 hearing in Tucson.[9] The smelters had been all-powerful until passage of the Clean Air and Water Acts of the 1970s. Now local people as well as the mainstream environmental groups had standing to show what they could do. Also important in stiffening enforcement were the acid rain issue—studies on the impact of sulfates on high mountain lakes showed that the smelters were the big contributors—and the effect of new regulations on visibility in class one sites like the Grand Canyon.[10]

In comparison with other air control issues, including power plants and automobile and urban smog pollution, the smelters were relatively simple to clean up. And yet the DRW lingered on as the nation's worst polluting plant until 1987, seventeen years after passage of the Clean Air Act. A large part of cleaning up the DRW had to do with implementation and enforcement issues, which will be discussed next, starting first with the Arizona SIP and the issue of intermittent control systems (SCS), followed by the so-called nonferrous smelting orders (NSOs) authorized by a 1977 amendment to the act that allowed certain copper smelters to apply for more time to achieve compliance, and concluding with the consent agreement that led to the final shutdown.[11]

Smoke control and dispersion programs like SCS were by far the preferred strategy of the smelters to deal with smoke in the early 1970s rather than constant control systems designed to capture almost all the sulfur dioxide and particulates. As the Arizona companies argued in petitioning to use them, "What is clearly needed is a reliable means of selective control, geared to changing meteorological conditions, that can be combined with a lesser degree of continuous sulfur elimination to insure compliance with ambient standards during adverse [meteorological] periods and yet avoid wasteful costs of unnecessary control during favorable periods." Tall stacks and SCS were not allowed under section 110 of the act, but the EPA initially was inclined to approve them as interim measures toward permanent compliance and with this goal in mind approved Georgia's SIP. This approach was successfully challenged in court by the Natural Resources Defense Fund in 1974, one of many suits brought against the EPA by environmental groups to stiffen the terms of enforcement.[12]

Furthermore, the emphasis on ambient air controls had the effect of encouraging the use of tall stacks and SCS by electric utilities which, as the largest contributors of SO_2, were now seen as the worst generators of acid rain, a newly recognized hazard. After 1970, over three hundred tall stacks were built, and the average height was increased from two hundred feet to over six hundred feet. "At these heights winds are stronger and more constant than they are closer to the ground. Consequently, more pollutants are staying aloft longer and traveling farther than before. The longer SO_2 and NO_x stay aloft, the greater is the likelihood that they will be converted into acids."[13] The upshot was that the first fruits of the Clean Air Act—intensified acid rain—were not encouraging, but as the effects of the long-range transport of pollutants became better known, the case for intermittent control systems was weakened. By the late 1970s, the use of tall stack was curtailed except for plants grandfathered in before passage of the Clean Air Act.

It was several years before cause and effect could be linked back to the utilities, but the smelters could not escape coming under review as contributors to long-range air pollution. Although the EPA's approach to smelter control was centered on ground-level pollution, the newly discovered risks of dispersion systems led to their curtailment in several court tests to section 110 of the act. The argument that prevailed was that tall stacks and SCS, together or singly, could be used to "fool" the network of ground monitoring stations being set in place around each plant to gather accurate emissions data for the public record. Data from these stations were gathered in 1973 before the smelters deployed SCS and were used in the ensuing arguments for enforcement of federal controls.

As well, the EPA's nonhealth-related secondary standard for SO_2 emissions by smelters was successfully challenged by Kennecott Copper Corporation in 1972. The court held that the evidentiary base being used to establish the annual arithmetic mean at sixty micrograms per cubic meter, which would become the secondary standard for sulfur oxide, was insufficient. The EPA did not challenge this ruling, and standards for SO_2 were revised in 1973. However, the agency "learned a great deal about how to survive judicial review in the ensuing years," defending its other standards and the methodologies used to set them against several more challenges from the smelters.[14] That the EPA was vulnerable to challenge on scientific grounds was not surprising given the history of smelter litigation. What was new was the federal government's commitment to further research as part of a holistic approach to abatement rather than the old plant-by-plant approach, as happened in the case of Trail. Nonetheless, the problem of specificity remained. Under the adversarial system, for both the EPA and the smelters it could be said of the ensuing rounds of ligation that the truth really was in establishing the details. In Arizona, for example, ten years of hard legal slogging lay ahead before the state's emission standards under which Douglas was forced to comply or shut down went into place in 1982.

It is against the background of legal challenge and scientific discoveries that the authorization scenario was played out between the EPA and the state of Arizona. Charged with developing a SIP for smelters, the original plan was submitted by the Arizona Department of Health Services on January 28, 1972, and disapproved in part on May 31 on the grounds that it did not contain either smelter SO_2 control strategies or regulations. The state's May 30 submittal that established regulation by a proportional rollback formula was also disallowed because it did not contain specific emission limits for each smelter to attain the NAAQS.[15]

Arizona nonetheless pressed on with a technology-based formula (linking abatement to the available technology, which would accommodate SCS) for attainment of standards and submitted it on January 7, 1977. This approach was disapproved also on the grounds that the test procedure would utilize a thirty-day average sulfur balance, insufficient, the EPA said, to attain the national air quality standards. The agency maintained that the averaging period for emission limits had to be on a short-term basis to protect both the three-hour and the twenty-four-hour NAAQS, whereas the proposed state smelter regulations prescribed technology-based "ultimate" emission limits. In effect, the EPA maintained in a letter to Governor Bruce Babbitt, "Arizona's monthly averaging period could permit violations of both short-term ambient standards for SO_2, without violating the State's emission limits." Emissions could be curtailed in order to deliberately mislead the monitors as to the actual tonnage being vented into the air.

Furthermore, the state was told that "SIPs must contain emission limitations requiring the use of constant controls sufficient to attain and maintain the . . . NAAQS."[16] This requirement meant that acid plants must be used in conjunction with the roaster and converter stages of smelting. A recent court decision *(Kennecott Copper Corp. vs EPA)* established that regulations for isolated sources of air pollution can allow SCS "on a temporary basis after installation of reasonable available control technology with the companion requirement that the ambient air quality standards ultimately are to be achieved through the use of positive control alone."[17] However, the court also specified that in addition to Kennecott's McGill, Nevada, smelter, five of Arizona's copper smelters must be allowed the opportunity to use SCS to meet air quality standards, including Douglas.

On January 4, 1978, the EPA promulgated what was intended to be the first effective SIP for smelters in Arizona, which would have required them to prepare for attainment in three years, by January 4, 1981. This requirement was contested by industry and the state's elected officials.

ASARCO, Inspiration, Kennecott, Magma, and Phelps Dodge all submitted petitions to halt the proposed EPA ruling. For its part, Phelps Dodge maintained that the company's three Arizona smelters (Morenci, Ajo, and Douglas) were subject to and had complied with the state's "final and binding regulations" that came into effect on December 22, 1976, and which the EPA was now contesting. The PD smelters were "presently operating under duly issued Operating Permits requiring compliance with all of Arizona's Air Pollution Control rules and regulations, including those requiring the maintenance of the ambient air standards for sulfur dioxide, . . . standards [which] are identical to the Federal Standards."[18]

At issue, however, was how these standards were to be met and measured. Did the goal of attaining a specified air quality require that actual emissions be *reduced* or could the environmental quality standard be achieved by *dispersing* over time or space, even in conjunction with permanent controls such as acid plants and stack precipitators? And what was the measure of compliance, EPA's more stringent or Arizona's less strict methodology?

Under pressure from the congressional delegation and Governor Babbitt, meetings were held in early 1978 with EPA staff to find common ground, if any, on the meaning of section 110, which disallowed SCS. The lawmakers were not satisfied. In a letter to Administrator Costle, Babbitt cited the EPA's "intransigence," whereas there should be room in section 110 "to give the State and EPA sufficient latitude to formulate a regulatory approach that will satisfy both of us." Babbitt prepared a court challenge to the ruling. (At that time, a former staffer recalled, Babbitt was not naturally drawn to this issue. "The plume, after all, had always been there since statehood."[19]) For his part Senator Dennis DeConcini wrote, "I had every confidence that the Arizona plan meets, basically, the intent of the Clean Air Act. . . . [However] I am disturbed that the economic future of Arizona could be determined, not as a result of the legislative process, but by the blind arrogance of a federal agency."

Pressure brought results in the form of a temporary suspension of EPA regulations on Arizona copper smelters while the state revised its SIP. "This," Babbitt wrote, "would give the state time to devise an emissions limit calculation technique which would be acceptable to both the State and the EPA." Costle's reply was conciliatory: "EPA certainly concurs in the view, expressed in your April 7 letter, that the rules we promulgated on January 4, 1978, do not represent the only possible way to meet the requirements of Section 110 of the Clean Air Act, and I share your optimism that our staffs can work out state implementation plan provisions for the Arizona copper smelters which will satisfy our joint obligations under the Act."[20] In May 1978, following Arizona's withdrawal of its SIP submittal, the EPA suspended publication of its ruling in the *Federal Register*.

Arizona's formula for achieving allowable emissions was now shifted to a multi-point rollback (MPR) approach, which was conditionally approved by the EPA in 1982 and finally approved in 1983. Under MPR, it was allowable to use the sulfur balance method for determining sulfur emissions. The new statistical technique allowed emission averages to be used instead of fixed emission ceilings. Industry wanted this alternative, because MPR was the more flexible approach, allowing for compliance under several sets of rolling averages rather than mandating an absolute limit on emissions at any one time. Variations in meteorology, concentrate composition, and smelting procedures, the actual ebb and flow of copper production, all were accommodated better under MPR. If the new standards were also somewhat relaxed, as researchers from the Environmental Defense Fund (EDF) pointed out—"these averaging techniques allow substantial increases in emissions when compared with the emission limits set by EPA in the 1970's"—nonetheless, the new Arizona SIP would still require drastic reductions from existing levels. Under MPR the state estimated that current average smelter emissions, already less than half of the level of the early 1970s, would be reduced by about two-thirds at the stipulated compliance date for each smelter.

However, instead of basing the emission standard on the maximum highest yearly reading, as the smelters wanted and the state was pushing, at the EPA's insistence the proportional reductions were based on hour/day/yearly averages of readings from the monitors. This stipulation was "the big breakthrough," Larry Bowerman recalled, achieved after more than a decade of hard struggle. Operating SCS to these standards would be much harder to achieve. To be sure, at Douglas the state monitored only in the United States, and this smelter still polluted freely into Mexico. "'All the other smelters kid about it. They wish they had an international border,' a state meteorologist said."[21]

The Arizona emissions target was set at 323,372 tons per year of SO_2. In 1984, the state's six active smelters were still emitting nearly 285,000 tons over the allowable limit; the uncontrolled DRW alone was spewing out 320,000 tons, followed by Magma's San Manuel plant, which was the second worst polluter, and both were operated under NSOs. Ironically, the older, less efficient plants were permitted to emit more, because costs to control pollution would have been higher than in such plants as Kennecott and Inspiration that had already installed constant controls. However, by 1988 the state's three remaining copper smelters were emitting only 176 tons of sulfur per day, a recovery rate of 88 percent.[22] By then the three Phelps Dodge smelters using the older reverberatory furnaces (Morenci, Douglas, and Ajo) were shut down, and Magma's San Manuel was well on the way to being modernized.

Meanwhile, the EPA began assessing fines against Morenci and Ajo for violations of the newly approved SO_2 NAAQS and was prepared to cite Douglas as well. Pressure for tighter standards mounted from the environmentalists. In comparison with the suspended federal ruling, emission targets under the new state formula were actually relaxed on all smelters except Magma and Morenci. Furthermore, the EPA had deferred its January 4, 1981, target date for compliance, an illegal act, as the Environmental Defense Fund's Robert Yuhnke pointed out. In March 1983 Richard Kamp challenged the EPA's January approval of Arizona's MPR SIP revision, and EDF filed a supporting brief. In 1985 the court upheld the EPA's approval, including the three-year date that had been set for the smelters to achieve compliance, which was January 14, 1986.

Yet it was another issue that really caught the attention of the EPA, now battling for survival in the hostile atmosphere of the early Reagan administration. Recent scientific work had verified the link between Arizona smelters and the long-range transport of smelter sulfates to the Rocky Mountains, which were causing acidification of high mountain lakes, a marker for acid rain. A highly effective way of linking the smelters, hitherto seen as local, site-specific control issues, to mounting national concerns about acid rain had been found.[23]

A final issue concerned Arizona's SIP for particulate matter, which the EPA also disallowed in 1973 but then approved one year later after revisions based on the process weight curve method were made. The smelters installed additional particulate control equipment. In 1975, Douglas installed electrostatic precipitators in its twin stacks for the first time. Then what the EPA referred to as "a complicated series of challenges" was made to its particulate matter SIP. In a 1975 lawsuit, Phelps Dodge and Magma Copper questioned whether the process weight regulations were obtainable. In January 1976 the Ninth Circuit Court remanded the case to the EPA for further study. And in response to a telegram from Senator Goldwater supporting this position, Administrator Russell Train replied, "You may be assured that we will not require compliance with the regulations by any copper smelter unless we are convinced that compliance can be reasonably achieved." The EPA agreed to suspend enforcement until a review of the standard was completed, a process that was still under review into the early 1980s.

With respect to Phelps Dodge, these challenges were withdrawn in the 1981 consent decree that set forth a compliance schedule for Morenci and Ajo, in which the parties agreed that the 1973 SIP on particulates applied.[24] However, in another twist in this complex litigation, the Ninth Circuit disallowed the voluntary dismissal on the grounds that it was no longer appropriate because of the court's January 10, 1976, remand to the EPA.[25] Not until 1984 was this part of the consent degree reinstated.

This issue of getting companies to recognize the state particulate standard was important for assessing penalties and it became a potential "deal breaker" in the shutdown discussions over Douglas. The EPA maintained that the period of forbearance ended in 1981 when PD entered into the consent decree agreeing to withdraw its challenge to the federally mandated particulate matter SIP. Phelps Dodge argued that the EPA's agreement with Goldwater to suspend penalties was still in effect, but recognition helped to moot it. Overall, the struggle over the particulate matter SIP was important in determining whether or not it could be used to backstop the health issue, that is, that the plume from Douglas was harmful to asthmatics. In the meantime, the EPA contracted with TRW to analyze the particulate problem in the six Arizona smelter towns. These detailed analyses were then used by the state to develop nonattainment area plans for these smelter sites.

Douglas so far had escaped being subjected to a mandatory plan for permanent controls, as had happened in the case of Morenci and Ajo, or having to pay the fines levied on the other two smelters for nonattainment of the SO_2 NAAQS. In 1986, DRW was still out of compliance in both particulate and SO_2 emissions, operating, as the EPA put it, "in chronic and gross violation of the applicable AZ SIP emission limitation for particulate matter" and emitting SO_2 "at levels far in excess of AZ SO_2 SIP emission limitations."[26]

With respect to SCS, both Phelps Dodge and Kennecott continued to use it in their old smelters at Douglas and McGill, the two holdouts in EPA's Region 9 without any plans in place for permanent controls, which they were legally empowered to do by Congress in its 1977 revision of the Clean Air Act. Under section 119 of the act, copper smelters could apply for NSOs in order to obtain short-term relief from emissions standards. The NSO provided for temporary deferral of compliance until January 1, 1983, with one renewal until January 1, 1988, under a second NSO if issued. The rationale for NSOs was that a downturn in copper prices during the late 1970s coupled with increased foreign competition, especially from Chile, was putting the industry at risk, so that a smelter which could not meet its ultimate emission limit because of technical or economic reasons should be allowed to stay open temporarily. With respect to Douglas and McGill, section 119 provided that "the interim constant control requirement [to be used with SCS] can be waived, if the smelter can demonstrate that installation of such controls would result in closure of the smelter. And if a waiver is granted, the smelter must attain NAAQS using tall stacks and SCS."[27]

However, section 119 allowed only a temporary easing of the regulatory vice. Upon expiration of the NSO the smelter had to comply with its ultimate emission limit under section 110 using continuous emission reduction technology. NSO regulations were published in 1980 and in a by now familiar pattern of litigation were soon tested in court. In 1982, the regulations were sent back to the EPA to correct procedural errors and for reinterpretation of "reasonable technology" as defined by a plant's ability to pay. For both Douglas and McGill it was argued that the cost of installing constant controls was uneconomic. Regulations curtailing the use of tall stacks, an issue of importance to the utilities and which were so important in the acid rain debate, survived a court test in 1985.

On June 19, 1980, Phelps Dodge applied for an NSO for Douglas with the special waiver. Arizona issued the NSO in 1982, but EPA's Air Division did not take official action on this request on the grounds that Arizona's SO_2 emission limits would not apply until 1983. Also, federal eligibility regulations for the NSO were under court challenge from the smelters. The political reality was that in the new Reagan administration mining interests and other groups were trying to weaken the Clean Air Act, which would have included amending section 119 to extend the NSO deferral option for five more years, to 1993. Senator Pete Dominici of New Mexico spearheaded this campaign. But because he was in the minority, the most effective advocate for extending the NSO was Congressman Morris "Mo" Udall, who was a strong environmentalist except when PD matters in his district were involved. However, the assault on the Clean Air Act crested in 1983 in the face of rising public alarm about acid rain. Congress, including most of the Arizona congressional delegation, responded by maintaining the air standards.

A second round of NSO regulations was promulgated in February 1985, and applications were submitted by Douglas and Magma's San Manuel, the state's worst smelter polluters. But behind the scenes, settlement negotiations with the EPA were begun at PD's behest.

By now copper prices were recovering, and with both Morenci and Ajo having been closed, the EPA expected an increase in emissions from Douglas smelter as concentrates were shifted to it from Phelps Dodge mines previously served by the other two smelters. Conditions were ripe for the EPA to close Douglas down: by a combination of penalties for continuing violations of particulate emissions and by what the agency called "an expensive set of production reducing demands we are making for the period prior to shutdown by January 1, 1988." As a condition for getting a second NSO, PD was willing to discuss a consent decree stipulating a plan for compliance before the shutdown, with the exception of penalties that it had successfully avoided for years, ever since 1975 when the EPA in discussions with Senator Goldwater had agreed to suspend enforcement of the particulate SIP. The maximum statutory penalty just for violating the particulate matter SIP, starting in September 1985 through December 31, 1987, the projected date of compliance, was a nontrivial $20.9 million.[28]

Two positions crystallized in the internal debates over penalties at the EPA. Region 9 supported the view that the fundamental goals of the agency—to guarantee shutdown by 1988 while providing interim controls for SO_2 and particulate matter to comply with the Clean Air Act to protect public health, to honor commitments to Mexico, and to remedy deficiencies in the NSO application—"should not be put at risk by inflexible adherence to the penalty policy." The Office of Enforcement and Compliance Monitoring back in Washington argued that "the precedent set by foregoing 'substantial penalties' in this case would be difficult to distinguish in other cases," and that Phelps Dodge had a lot to lose by not negotiating a consent decree, indicating that their present position was likely a bluff. Thus, "even if negotiations were to break down on penalties, litigation could still realize the agency's goals."[29] Furthermore, the fine had to be large enough to remove any economic benefit resulting from failure to comply with the law—controlled smelters were being put at an unfair competitive advantage by Douglas—while providing deterrence to potential violators. The agency's starting position, put to PD on April 30, 1986, was $7 million, a smaller figure that was substantially reduced even further by the time of final settlement.[30]

While the long struggle between the EPA, Phelps Dodge, and the other operators was reaching closure, conditions were also ripe for an all-out effort by the NGOs to close the smelter down. By 1984 this effort was in high gear. Through such means as networking, bringing suit, testifying at public hearings, buttonholing officials, and using targeted mailings and well-timed publicity, their knowledge of the issues and attention to detail made a difference. For example, the Environmental Defense Fund's Robert E. Yuhnke had filed a brief in support of Richard Kamp's suit (on March 14, 1983) against the EPA for approving the Arizona MPR SIP revisions; he filed suit against Douglas (on November 29, 1984) for failure to comply with various requirements of the Arizona MPR regulations on SO_2 emissions; and he notified the EPA administrator (in a letter dated April 9, 1985) that EDF would challenge the agency's second-round NSO regulations. Kamp, an environmental activist operating out of Bisbee, had founded the Cochise Smelter Study Group (later known as the Border Ecology Project) with Michael Gregory in late 1979. They soon joined forces with EDF, a large national organization, and with the Southwest Environmental Service, a Tucson NGO led by Priscilla Robinson, a former lobbyist for Planned Parenthood who had become known for her work on land use and water issues.

The work of this triumvirate was important. For example, their organizations took the lead in mobilizing Arizona public opinion against the smelters while inducing the EPA to bring the health and Mexican linkages to the table for the first time, as will soon be discussed. Kamp, a self-described "Chicago dropout" from an engineering family, was adept at ferreting out actual practices from smelter personnel and at passing and generating information. By the time Douglas shut down, he had become something of a folk hero among U.S. environmentalists. Yuhnke, a skilled environmental litigator, was a premier example of EDF's dogged commitment to making the system work better. And Robinson, whom Babbitt later dubbed "the Den Mother of Arizona environmentalists," did her homework on air issues and the smelters, built networks, testified several times before Congress, and for awhile when the smelter issue was hot appeared on ABC's *Nightline*, was covered by the *New York Times*, and wrote frequently for the *Arizona Daily Star*. Taken singly and, even more so, collectively, the team was an American phenomenon and justified the fondest hopes of environmental leaders in the Congress like Senator Muskie and others who had built public participation into the Clean Air Act.

With respect to citizen participation, a signal advance of the 1970 Clean Air Act was the standing given to public comment, the capacity to sue the U.S. government for failure to enforce, and citizens' right to know, including access to information that by law had to be made public. For example, access to monitoring data was critical to the struggle with the smelters. In retrospect, it was the use made of these "bricks and mortar" rights, rather than the demonstrated capacity to "throw bricks" on the part of groups like Greenpeace, the stunt meister nonpareil, that made the bigger contribution in bringing the smelters under control.

Long used to winning, PD lawyers seemed baffled by the mix of economic change and demographics that was transforming the culture of their state. Many of the Arizona environmentalists were in fact newcomers. Some, like the radical feminists who led the new Group Against Smelter Pollution (GASP) in Bisbee were single-issue people: shut Douglas now! GASP, whose petitions included the names of very few Hispanics, mistrusted Yuhnke's careful, legalistic approach, and for his part Yuhnke was baffled when they took him on.[31] A phlegmatic Pennsylvanian who was very much a mainstream environmentalist, Yuhnke did the legal work for EDF from his regional office in Boulder, known as "the People's Republic of Boulder" for its avant-garde politics and alternative lifestyles. This new social mix had spilled over into Arizona by the time EDF got involved in the Douglas case.

It was a world apart from the old mine and smelter towns where my grandparents lived and served Phelps Dodge, including Douglas, Gilbert Davis's last posting. Nancy Wrona, a former staffer for Governor Babbitt on environmental issues, recalled the 1982 state hearings on the NSO applications from Douglas and San Manuel. In the audience was a young mother breast-feeding her child. Wrona recalls "the look of absolute astonishment on the company attorney's face when he looked up from the hearing table and saw her. They were at least three, maybe four cultural generations apart."[32]

The new world also included health seekers and retirees, many of whom had recently moved to Arizona, including Cochise County and Douglas, of all places. Concerns about health had surfaced for the first time at the 1982 hearings. Letters sent to the EPA on the eve of the May 1986 public hearings held in Douglas on whether to renew the plant's NSO operating permit included many comments from asthmatics and physicians. The tone was that of people who were used to being heard and able to defend themselves. For example, Donald H. Henderson, a recently arrived retiree in Sierra Vista, had worked for many years for an international chemical company. Both he and his wife suffered nasal problems because of the plume. He wrote:

> I was in Douglas the evening of April 8, 1986, and noted that PD was in full production with dense clouds being emitted from the stacks. I ask you to respond officially as to why full production is turned on at night. I await your response. After spending many years as an executive for an international chemical company I know the *true* answer. When it is dark you run wide open and hope by morning the emissions as well as the effluent are dispersed so that you cannot be caught. I also know how the chemical and the steel industries threaten to shut down so that the politicians will close their eyes to avoid plant shutdowns. My company cleaned up.

In conclusion, he wrote with conviction: "Their attitude very bluntly is the public be damned. The days for accepting that outlook by big business are long gone. Make them abide by the law just as we citizens should!"[33]

Among the many letters was an appeal to shut the smelter down from Robert J. Wick, the owner of Wick Communications Company and a transplanted Ohioan who ran a chain of small-town newspapers, mostly in the West, from his Sierra Vista offices: "I hope that your agency will truly fulfill its reason for being which is to protect the environment and the people who have to live in it. Other companies have had to comply and did. PD believes itself to be outside the law and it is not fair to the people of Arizona." Looking back on this period, Wick recalls days when the mountains fifteen miles away from Sierra Vista would be obscured by smoke blowing in from Douglas. He was worried that southeast Arizona would not grow and be developed because of smoke pollution. With the prospect of Mexican smelters coming on line in addition to Douglas, the area could become instead "a hell hole of pollution."[34]

But the old world was still there, as revealed in petitions from Douglas smelter workers, many of whom had Spanish surnames, asking the EPA to allow DRW to continue operations, and letters from several old-timers including one from Marguerite R. Askins of Bisbee who said:

> It is difficult to understand the hullabaloo over the smoke from the PDC smelter at Douglas.
> My husband's family moved to Douglas in 1902 and I, a willing transplant from upper NY State, have lived in Cochise County for almost 30 years. The stacks at Douglas have stood tall since the early

1900s, and if the smoke is as bad as claimed, why have so many people moved to this area? I can sympathize with the asthmatics who feel that the smoke is aggravating their conditions . . . [but] why not move on? Why put 354 Douglas families and 34 Bisbee families out of work? Unemployment is also hazardous to one's health!

Moreover, she concluded thoughtfully, "More people favor the continued operation of the smelter than oppose it. Opponents are naive if they think the shutting down of the Douglas smelter will force Mexico to install controls on their smelters."[35]

A Bisbee contractor was less restrained: "I have lived in Arizona and worked around copper smelters for 50 years" and am in excellent health, so "why is it people come to Arizona and immediately blame our copper industry for those such problems? If it weren't for our copper industry, Arizona would still be roamed by the Indians." But "now we have the EPA, out-of-state environmentalists, and a governor seeking publicity trying to shut down the smelter because it is bad for our health." They should leave the state for good, he fulminated. "The best thing would be to load up Babbitt, Mr. Kamp, and the rest of their hippie environmentalists in a bunch of cattle trucks and take them to Columbus, New Mexico, like the old-timers did with the Wobblies [in 1917]. The Wobblies tried to tear this country down, and so are they and the sooner we ship them out the better."[36]

This brief summary of a very large file gives a flavor of the issues and forces involved: jobs counterpoised with concerns over health and the Mexico linkage; old Arizona and the newcomers; ethnic relations; and the long arm of the federal government reaching out to a once dominant corporation that by 1986 had returned to economic health but was very much on the defensive. An early 1986 poll taken by the Douglas *Daily Dispatch* found the community "pretty evenly split on the issues involving Phelps Dodge Corp.'s copper smelter . . . [although] the highest percentage figure in any response does not want the smelter closed down now." When asked whether visitors and potential new residents found the smelter pollution less desirable than pollution in Tucson and Phoenix, respondents said yes by a small margin but tilted the other way when asked if SO_2 from the smelter was hazardous to the health of Douglas residents—even those without respiratory problems.[37]

That the issues of health and the changing economic base of Douglas were now very much on the public agenda owed much to the environmental activists. Babbitt himself embraced these issues in an op-ed column in the *Los Angeles Times* in which he came out publicly for the first time against extending the smelter's operating permit. "The copper smelters pose not only health and environmental threats, but also an economic threat," Babbitt said. "Southeast Arizona is one of the most desirable places for tourism, retirement and light industry, but that development can't take place."[38]

If public opinion was still divided on these issues, the fact that the public was now engaged was part of the contribution of the environmentalists in shaping the issue of smelter control. Their role will now be examined, with close attention to the triumvirate of Robinson, Yuhnke, and Kamp. I will begin by taking a close look at the activities of the Southwest Environmental Service (SES), run from 1975 to 1988 by the Tucson activist Priscilla Robinson.

Initially established to encourage citizen participation in land use planning and water quality issues, SES, in response to the 1977 amendments to the Clean Air Act that introduced section 119 and the NSOs for smelters like Douglas, shifted to issues of urban and rural air quality, with emphasis on smelter emission control. To encourage citizen participation in the planning process for the Arizona SIPs, SES and other groups secured EPA funding to hold a Clean Air Workshop in 1978. The EPA then reprinted the handbook Robinson prepared entitled *Blue Skies: An Arizonan's Guide to Clean Air*. SES became actively involved in monitoring the NSOs for the Douglas and San Manuel smelters and helped to establish the Mexican linkage. Robinson was very effective in lobbying Congress to turn back the legislation that would have weakened the Clean Air Act and would have extended the NSO deferrals until 1993. She attended meetings of the citizens' advisory committee on revising Arizona's SIPs. Under her direction, SES joined suits to bring Phelps Dodge and Magma facilities into compliance with the Clean Air Act when they applied for second-round NSOs in 1985. Using the legislative and legal processes to create and enforce state and federal laws was the hallmark of the SES approach.[39]

However, shutting Douglas down turned out to be the group's last big project, and it disbanded in 1988. Robinson went on to pursue other interests and, following a time-honored American tradition, put what she had learned about mining to use, especially her smelter contacts, as a consultant to the industry. In this there was nothing untoward: she was never antimining. In the course of her smelter work, Robinson had developed good working relations with men from ASARCO and Kennecott who broke ranks with Phelps Dodge, considering that company's intransigence as both out-of-date and as a cover for putting them at a competitive disadvantage. That Douglas had a competitive edge over smelters with controls was an argument Robinson included in her congressional testimony against extending the NSO provisions until 1993. After testifying in December 1981 she made good industry contacts and learned a lot from them. As she wrote the SES board, site visits to ASARCO's new facility at Hayden and additional research confirmed for her "the validity of my general approach to smelter regulation, which is that smelters need to modernize their processes (à la ASARCO) in order to remain competitive with foreign industry, and that federal clean air laws should be designed to encourage this kind of modernization."[40]

In writing this book, I was fortunate to secure from Robinson a nearly complete file of the monthly reports that she prepared for her board on SES activities and her written comments in the form of a yearly chronology of events. When read in conjunction with the SES archive, this source provides a very good idea of the anti-Douglas campaign from the perspective of a leading player among the NGOs.

To be sure, it is one perspective. Consider the "Rashomon" effect. Saving the company through a drastic restructuring was how Phelps Dodge saw its actions. Wearing the industry down with court-tested standards and then tightening the regulatory noose on the polluters were how the EPA saw its mission. Stiffening the resolve of a reactive and slow-to-act agency (the EPA) while endeavoring to make sure Phelps Dodge had no friends left in the governor's office or Congress was how the antismelter triumvirate saw its role. On balance, historical analysis of the play of events must include elements of all three perspectives. Here follows Robinson's own chronology of key events:

1978: In late 1977 SES becomes interested in smelters through its concern with visibility plans and class one areas like the Grand Canyon and national monuments. It concentrates on educating environmental activists about the law through the workshop held in June and the organization of working groups. Research on *Blue Skies* begins.

1979: Learning that the EPA was planning to hold hearings in Tucson on the NSO regulations, Robinson "realizes I would have to do something about them, as no one else knew anything about them either including the National Clean Air Coalition (NCAC)," which was slow to embrace the smelter issue. Dick Kamp contacts SES in December, and the Bisbee group organized themselves "and appeared by magic, from my point of view."

1980: SES focuses on visibility issues, with Kamp learning about smelter technology and taking pictures of the Douglas plume, while Robinson serves on the Four Corners Study Group for the NCAC and learns about modeling, visibility measures, particle transport, and "lots more." In July, Kamp and Robinson go to Salt Lake to testify on the EPA's proposed visibility regs.

1981: SES learns that the Reagan administration is planning to gut the Clean Air Act and extend NSOs until 1993. In March Robinson meets EDF's Bob Yuhnke at an NCAC event in Washington. "He was pointed out to me as the only person there with any interest in, or knowledge of, smelters," and they agree to work together. On June 5 she testifies before

Senator Stafford's Committee on Environmental and Public Works and says the section 119 waiver of continuous controls under NSOs is not working. Using Kamp's data and findings, Robinson shows that SCS at Douglas was not controlling emissions: "The whole PD monitoring plan was a farce, and they did indeed turn up the plume at night into Mexico." In July the administration plan is leaked, and SES starts a media campaign. "We got several reporters hooked on smelters during the year, including the *Star*'s Jane Kay, who would later win a Nieman Award for her articles on smelters."

That fall Robinson begins to work with Governor Babbitt, who had appointed her to the state parks board. On December 14, she testifies before Representative Waxman's Subcommittee on Health and Environment, again on section 119, and puts EDF research on the long-range transport of smelter sulfates to high mountain lakes into the record. She uses an economic analysis supplied by Yuhnke to support the view that uncontrolled smelters have an unfair competitive advantage over those that are cleaning up. In talks with the Arizona delegation and others during this trip, Robinson learns that the national NGOs, particularly the Sierra Club, want to protect the act but are reluctant to oppose section 119 for fear of displeasing Congressman Mo Udall, whom they want for the Alaska wilderness legislation. (Udall, the congressman for Tucson, is testy and short with Robinson and the Bisbee group when they catch him at the airport.) [A great environmentalist, Mo, who died in 1998, took care of copper interests in his congressional home base.] The president of Inspiration Copper company says that "your presentation [before Waxman] was balanced and well informed," and during this trip she begins to work with representatives from Inspiration and Kennecott.

1982: The year begins as Greenpeace arrives out of the blue, informing Robinson that they are going to climb the stack at Douglas as a media event. While not associating herself with this action, Robinson advises them to climb the San Manuel stack instead because Douglas was remote from the press and because the law enforcement there under Sheriff Jimmy Judd was still in a Wild West mode. Mindful of her contacts with Babbitt and the congressional delegation, Robinson does not want responsibility for the stunt, which occurs shortly before the EPA hearings on NSO applications at Douglas and San Manuel. "The Greenpeace team stayed around and testified at the hearings, in both San Manuel and Douglas, which livened things up and attracted lots of press." For these hearings SES is getting help from Yuhnke on the NSO regulations, but the EPA would not approve the NSOs. "EPA regarded section 119 and the administration position on the extension as a virtual waiver of everything for smelters. This is why they never bothered to review the NSO applications, etc. We were headed into a legal morass." By year's end the "smelters were regarded as a major villain by the press. I could get an editorial out in the *Arizona Daily Star* virtually for the asking. This was much resented by the Udall staff." Robinson, Yuhnke, and Kamp now see themselves as a team. SES continues as a source of information and as an advocate for decision-making, but now, she reminds her board, the primary focus is on groundwater regulations being developed by ADHS and the Clean Air Act amendments.

1983: The National Clean Air Coalition finally takes a firm stand against any extension of the exemption for smelters beyond the present limits of December 1987, and Robinson in meetings takes "the position that any smelter amendments to the bill should tend to encourage modernization, such as ASARCO is undertaking, and should not give a competitive advantage to smelters that delay. While no one wants to close down smelters while the industry is depressed," she informs the board in her report for April, "benefits should not be tilted toward the dirtiest smelters." After redistricting, Jim McNulty becomes the congressman for the Douglas area and with his strong union ties is much easier to work with than Udall. On May 20, he and Senator DeConcini hold a hearing in Tucson on the proposed World Bank/International Finance Corporation loan to a Mexican copper company to expand the mining and uncontrolled smelter complex in Cananea.

The hearing on foreign competition and the Arizona copper industry is "a big love-in," bringing together unions, environmentalists, and industry just before the strike. McNulty appreciates Robinson's hard work in making this event go well. In July the Morenci strike begins. "The strike in July changed everything and eventually made it possible for us to win," Robinson recalls. Babbitt was already planning to run for president, and he "bitterly resented the position that PD put him in, in having to call out the National Guard, thereby forever winning union anger. He would have done anything to hurt PD after that, and anything to win union favor." With their strong union ties, both McNulty and DeConcini are on board, but only Udall holds out for another year. A second round of hearings on NSO regs is held in Phoenix in October, and Yuhnke finally visits the state at Robinson's insistence. (The EPA did not approve these regs "until driven by our lawsuit in 1985.")

1984: "In retrospect, 1984 was the turning point year. We moved from the defensive to the offensive." In Washington, politicians are moving away from the NSO extension that loses support because of two factors: (1) "the strike fallout, which was very damaging to PD politically and we used every bit of it," and (2) "the Nacozari smelter issue [covered in the next chapter]." Yuhnke and Michael Oppenheimer of EDF have their western acid rain road show going by then, and it appears in several forms before the final publication by EDF and eventually in *Science*.[41]

Babbitt sees the Yuhnke slide show. After seeing this presentation on the acidification of mountain lakes, Udall, in a June meeting with Yuhnke and Robinson, comes out against extension.

"Yuhnke was the star at a western acid rain meeting in Gunnison in July. I had a chance to really talk with him, and he explained that we were finally ready to file a lawsuit.[42] I was delirious with joy." Yuhnke files the suit in federal court in Tucson in November, "and he and I held press conferences in Tucson and Phoenix, then flew to Chino, N.M., for dinner with the top management of the Kennecott smelter and mine" (which Phelps Dodge subsequently purchased in 1987). The upshot, she concluded, was that "things were really moving our way. Politicians all on our side now, industry support for closing Douglas, acid rain stuff working, Dick's project paying off, and the legal action started."

1985: This year the chief objective is linking the two issues of the Phelps Dodge operating permit and the Mexican connection. Kamp orchestrates the Mexican issue as a matter of helping the copper industry so that Congressmen Kolbe and DeConcini are locked into closing Douglas as part of the deal. Kolbe, a former SES board member, is really an environmentalist at heart, but he is also a pragmatist who towed the administration line for awhile. The triumvirate testifies at a Senate hearing on acid rain in Denver in August and at a House hearing in Albuquerque earlier. "We had a nice line of coordinated testimony going and spent a lot of time together. Between EDF's western acid rain stuff, the strike, the Mexican connection, etc., we were national news." As well, Robinson recalls, "my main job was keeping the pressure on Babbitt from all sides, especially the media."

During her Washington trip in June she informs and updates legislation on western acid rain, but Mo Udall "could not be convinced of the need to add western acid rain to his existing eastern acid rain bill he is introducing in Congress." However, in September, "I *finally* got Udall [who was now quite ill with Parkinson's disease] to agree that Douglas would have to close as part of the international deal." Lee Lockie, head of the Arizona Department of Health Services, "had been turned around and was working to do the right thing on the operating permit decision." In October a southern Arizona coalition of environmentalists, including SES, GASP, and EDF, individuals including Bob Wick, physicians, businesses, and the United Steelworkers petition for standing in the Douglas permit renewal hearings before the Arizona Department of Health Services. In seeking standing, Yuhnke represents Robinson at these hearings and lays the groundwork for reopening the lawsuit against PD. Meanwhile, the SES efforts to prevent groundwater contamination bring on a fight with the mines, agriculture, and the secretary of commerce, among the most powerful groups in the state.[43] "This year the team works closely on all parts of the effort—state, local, national, international, press, politics, etc."

1986: This is Bob Yuhnke's year "to play all the legal cards in the endgame. Dick and I do not always follow the action." He orchestrates the asthma issue (see below). Since May 1985, the EPA has been granting temporary stays of enforcement of the SIPs and federal regulations while they review the NSO. EDF sends a letter of intent to sue—"this action should have the effect of forcing EPA to act on the NSO"—and Governor Babbitt sends a letter to the EPA asking them to act soon and to consider denial. On May 17 a hearing is held at Cochise College in Douglas on the EPA's proposal to deny the Douglas NSO. On July 10 the smelter closes. The EPA, Phelps Dodge, and the state reach agreement on the smelter, which will reopen but must close on January 15, 1987, under terms of a consent decree that is lodged in federal court on July 29. However, EDF's 1984 suit is still active. Kamp and Robinson are parties to the consent decree, under which Douglas closes one year earlier than potentially allowed under the law. "It was all over but the party."

1987: January begins with a victory celebration in Bisbee, which is covered in fresh snow. The party is quite bizarre. Yuhnke is going with a girl who is into alternative lifestyles, and her house is filled with strange art. Everyone is drinking beer and smoking pot when Robinson and Yuhnke arrive late because of the storm. Robinson asks for some hard liquor, and, amazingly enough, a bottle of Jack Daniels is produced for this fifties-something activist from another generation. The three principals then drive to Douglas for pictures at the deactivated smelter. The Magma wrap-up continues, with Kamp and Robinson granted intervenor status and enforcement powers.

The final smelter entry in Robinson's reports to the SES board is August 20, 1987. The upshot is that "PD has no operating old smelters in Arizona—but has two modern smelters in New Mexico. One is the compliance smelter they built at Playas in the 70's and the other [(Chino) was] purchased from Kennecott in 1987, the one Yuhnke and I toured the day after they filed the lawsuit, in November 1984." There are three smelters operating in Arizona now, she wrote the author in 1988: Inspiration, built in the early 1970s, a state-of-the-art smelter that never worked very well; ASARCO's Hayden plant that was modernized in 1982 using a new INCO furnace; and San Manuel, which has completed modernization with flash-furnace technology.

This concludes my short summary of key events, told as faithfully as possible through the voice and point of view of a leading environmentalist. From first engagement, to networking, to congressional testimony and working with public officials, to the final shutdown, Robinson's analysis and chronology of the flow of business provide an excellent snapshot of an environmental NGO in action. Of all she did, perhaps Priscilla Robinson's single most important contribution was the successful effort to prevent the Reagan administration from extending the NSO waiver of continuous smelter controls under section 119 for five more years (until 1993). Or when all is said and done, perhaps it was in furthering citizen education, SES's core mission, and then showing what a well-prepared and well-networked citizen activist could do.

Reflecting on her role Robinson said, "I spent thirteen years working for solutions and was never anti-mining; I just wanted to stop that smelter."[44] "She was really influential with the governor and with the congressional delegation," Nancy Wrona told me shortly after these events transpired. The work with labor, public officials, the environmentalists—each had a strategy, which began to interface. "Priscilla was the key to building this web." In looking back on these events, Wrona felt that from her perspective in the governor's office, Robinson was the one who really shut down the plant under section 119 of the Clean Air Act. "Babbitt loved her," Kamp told me later. "Dick Kamp and Bob Wick and Priscilla Robinson did a lot," Babbitt recalled. "They really got me going on this one, and Priscilla kept me locked in. Only in America could something like this happen."[45] And it is most certainly a credit to the Republic that people like Priscilla Robinson have the opportunity and the public space to act. To be sure, her effort was also a partnership, and the roles and contributions of Yuhnke and Kamp should now be assessed.

The EDF's Bob Yuhnke's behind-the-scenes role was only apparent to most people when he started showing up at smelter hearings in 1984. His work on the Douglas smelter should be seen first in the context of the acid rain issue, the scientific and legal aspects of which have long been a major concern at EDF.

As mentioned, Yuhnke began collaborating with Priscilla Robinson in early 1981, and EDF's report entitled "'Acid Rain' Research in the Intermountain West" was entered into the record with her testimony before the Waxman subcommittee in November of that year. Yuhnke helped prepare the report, which found that the increases of acid deposition into high mountain lakes that were being discovered by researchers probably resulted, in the West, from the long-range transport of NO_x from automobiles and SO_2 from smelters and power plants. Since current SO_2 emissions from smelters exceeded 1.1 million tons per year, and some 80 percent of the sulfur dioxide emissions from major sources in the West was produced by the smelters, this level was far above the less than 400,000 tons allowable under current state implementation plans. It followed that these reductions could be achieved only if the compliance exemptions enacted in 1977 under section 119 were repealed.[46] Fortunately for the cause of smelter control, the momentum to extend the NSO control exemptions through 1993 did not prevail. What is important to note here is that the linkage between smelters and acid rain in high mountain lakes had been raised. It was now incumbent on researchers to prove it.

With the exception of EDF, the main national organizations working on acid rain were fixated on power plants well east of the Rockies, specifically the reliance on high-sulfur eastern coal to fire midwestern generating stations. The NCAC only came to Arizona late in the game, to Robinson's frustration. Western resource industries were quicker to respond to the implications of the research being done by EDF and others on high mountain lakes. Companies producing low-sulfur coal in Wyoming and Montana stood to benefit from the imposition of controls. Oil and gas producers worried about emissions limits and saw their industry at risk, particularly the plans for an EXXON natural gas sweetening plant and a Chevron phosphate project, both of which were slated to produce significant amounts of SO_2 when they came on line in Wyoming. In January, Yuhnke received a request to meet with western energy development interests to discuss a possible compromise on NSO extensions.[47]

In fact, EDF's research to establish the smelter link was done in conjunction with the workup of emissions data from Wyoming refineries. Data on sulfate precipitation measured weekly by the National Atmospheric Agency stations on mountain lakes in Colorado were found to correlate nicely with smelter emissions. When these decreased substantially during shutdowns in the early 1980s, the monitoring stations recorded less acid deposition. The results of this research were duly published later on in *Science*, which helped to validate the methodology. In a study commissioned by Phelps Dodge it was argued that because of regional differences, there was "significantly less potential for acid rain deposition in the West, particularly in the more arid and semi-arid regions of the Southwest." In fact, this company scientist said, "large areas [of the West] probably would benefit from acid deposition."[48] However, the implications of EDF's ongoing research were not lost on industry and the regional politicians involved. Matters came to a head in the summer of 1984.

Prompted by the granting of new source permits for the Wyoming energy projects, readings from Wyoming monitors were worked up in June. Readings from the five Wyoming stations went up and down together, and when the copper emissions were overlaid on top of that curve, a tight fit was obtained. Knowledge that Mexico was planning to build a very large and possibly uncontrolled copper smelter at Nacozari across from Douglas in Sonora now played neatly into the acid rain dynamics, which meant bad news for the Rocky Mountain energy industry. Without tighter controls on the U.S. and Mexican smelters, there would be no cushion to support some development of the gas sweetening, synfuel, and coal-fired power plants being planned for the West in the years ahead.[49]

Yuhnke recalled how that message was received: "Given the current levels of acid deposition, you can get new plants in Wyoming, but add the Nacozari plant and you will get large changes in lake chemistry. Therefore, the Wyoming plants shouldn't be built." This point was an economic bombshell in Wyoming, which had been banking its development future on the old 1970s argument that an oil and gas shortage existed. The governor set up a committee to investigate acid rain. Although the permits were issued, the ensuing furor gave the EDF team direct access to EPA administrator William Ruckelshaus, whom they briefed for one and a half hours in October.

Brought back to run an agency still reeling from the shambles created by the antienvironment extremist Anne Gorsuch, Ruckelshaus could see a serious problem looming and was keenly interested in their presentation, which linked the smelters with the lakes and with Mexico. As a result of EDF's actions, the western smelter issue had already been linked to the larger concerns over acid rain; now as it was becoming internationalized, the issue became what to do about Mexico. As Yuhnke relates, Ruckelshaus was driven to action by the specter of a new acid rain problem in the West, with an additional set of demands being created just as the eastern interests were hitting him. How would he explain how a new acid rain problem had come up on his watch? With the Canadians pressing hard on the acid rain issue, Ruckelshaus did not want to fight on two fronts. But the United States could not hold Mexico to smelter controls if they were lax here.

A way to solve this problem was worked out in the next few weeks, as will be discussed below. What is important to relate here is that Ruckelshaus suggested a framework that now informed the strategy of EDF in its commitment to solve both the Douglas and San Manuel NSO extension problems. As Yuhnke recalls, he said that "we couldn't do it on the basis of acid rain, for this would stimulate an adverse reaction from OMB, thus provoking the government while complicating State's relations with Mexico. And since this approach would bear on the Canadian acid rain negotiations, he didn't want to give Canada a stick to beat on the U.S. So his idea was to do it as a health issue."[50]

Yuhnke then proceeded to play the health card. EDF had already been investigating the health effects of fugitive dust associated with coal mining, uranium milling and mining, and spent shale disposal from oil shale, a good Rocky Mountain issue.[51] That its reporting on sulfur-producing sources and acid rain was now hot owed much to linking the smelters to the threat to mountain lakes. This approach grabbed the attention of industry and government, and the regional press picked it up

Having long been interested in the visibility standards in the Clean Air Act, health concerns, which the primary emission standards for SO_2 and particulates were designed to protect, were a good issue for EDF.[52] While continuing to testify on acid rain,[53] Yuhnke was now ready to litigate the Douglas operating permit and to develop the case for health based on the same deliberate blending of scientific proof with a good litigation strategy and publicity.

On August 1, 1985, the Arizona Department of Health Services (ADHS) received a petition from EDF, Groups Against Smelter Pollution, and the United Steelworkers of America, which claimed that sulfur dioxide levels in the Douglas area constituted an "imminent and substantial endangerment to public health" and called on ADHS to either require Douglas to prevent short-term peak exceedences at a level of 1,000 ug/m³ or to revoke the state air quality operating permit. This action would have required a more stringent standard than the existing NAAQS. (The current standards were 0.14 ppm for twenty-four hours [primary] and 0.5 ppm for three hours [secondary].) When the state refused to consider adding the more stringent standard, on August 23 the EPA was similarly petitioned. Using data from hourly readings from state and company monitors over a five-year period, it was found that exceedences of more than 1,300 ug/m³ for one hour, or 0.5 ppm, had occurred with sufficient frequency in three smelter towns to cause distress to asthmatics. The issue was one of "endangerment to health" and thus actionable under the federal and state standards that limit SO_2 concentrations in ambient air to 1,300 ug/m³ during any three-hour period. And since Douglas "has had a history of failure to operate its SCS so as to prevent exceedences of standards," a margin of safety should be built in below this standard in order "to protect asthmatics from ambient exposures in excess of concentrations that would cause adverse effects to asthmatics."[54]

In October the state conducted its own survey and found that DRW was in fact responsible for several exceedences during very short periods. ADHS's Lee Lockie took the position that these very high short-term concentrations would have to be addressed in the final NSO permit. Her agency noted that by far the greatest number of complaints were coming from Douglas, not the other smelter towns. This disparity might indicate that "the potential threat to public health may be greater in Douglas," but in order to complete its investigation the agency would need to gather a great deal more data on asthmatics.[55]

Whether or not smelter smoke actually was harmful to asthmatics was difficult to pin down, especially in the face of contradictory claims by health sufferers and physicians made in affidavits obtained by Phelps Dodge and by EDF. For example, a physician testifying for Phelps Dodge referred to a study of death rates among smelter workers that did not reveal any overall mortality excesses when compared with the mortality of the total U.S. population. However, as Michael Gregory pointed out in his testimony, the same study showed an association between exposure to SO_2 and deaths from emphysema. (However, the emphysema study was not correlated for cigarette smoking.)[56]

As for the asthmatics' argument, it was bolstered by testimony presenting what had been learned about the effects of even small concentrations on asthmatics and by the testimony of a credible and telegenic sufferer in the person of June Hewitt. Her videotaped account of a trip made from her ranch into Douglas, showing the effects of exposure to the plume (as captured on the monitors), was brilliantly done and compelling. The tape was widely circulated and received favorable press. This true-life performance made the case: "Asthmatics like Ms. Hewitt should not have to expose themselves to the risk of a severe asthma attack as the price of undertaking the daily routines of working on a ranch and traveling into town for feed and supplies. It is this kind of health hazard that the Clean Air Act was intended to prevent," Yuhnke argued.[57]

Until then, the health issue had been raised frequently (especially at the 1982 Douglas and San Manuel permit hearings) but had not been well developed. In truth, state health officials had been reluctant to play a strong role. However, the data on the SO_2 effects on asthmatics provided by the environmentalists "pushed people over at the eleventh hour. It was the asthmatic study that got them the good relationship with the governor," Nancy Wrona recalled, "so they could pick up the phone and call him, developing a useful relationship."[58] Yuhnke himself "always felt it was the key to moving politicians," Priscilla Robinson said, but she herself felt that "the key ones had already been moved by us, by the strike and the Mexican affair." The real impact was on the EPA, she recalled, which did not want to deal with the health issue at all: if they shut down Douglas, the issue would go away. Of course, her observation was from the NGO perspective.

To be sure, Region 9 officials had been talking with their state counterparts on the effects of SO_2 for some time and saw interest in the smelters rising rapidly as ADHS took on more staff in the early 1980s. In 1982, EPA's Office of Air Quality Planning and Standards recommended that another primary standard (0.25–0.75 ppm for one hour) be adopted to protect sensitive populations.[59] In fact, the health issue and specifically the emphasis on short-term peak emissions tied in quite well with the EPA's long campaign to get rid of SCS, which was still allowed as a stopgap control measure under the NSO process. "SCS could not react fast enough to handle these short-term peaks, and this was one of our strongest weapons," an EPA engineer recalled.[60]

What is beyond question is Robinson's assessment that Yuhnke "pu a huge amount of effort into the asthma issue, and the asthmatics were also very important to us—the ones who testified and were willing to be deposed by PD attorneys."[61] In April, notice of the EPA's decision to deny a renewal of the Douglas NSO appeared in the *Federal Register*. Health was referenced, but SCS, Mexico, and the federal/state air standards were given much more prominence in the argument for denial.[62] In the subsequent public hearing held in May 1986 at Cochise Community College in Douglas, the weight of testimony, based largely on the health and local economic issues, ran heavily against Phelps Dodge.

Drawing attention to the overall record of compliance, PD's Pat Scanlon said that the company "can and will satisfy all of EPA concerns" over technical deficiencies in the SCS system and fugitive emissions. He also claimed that the newest SCS system would in fact capture all short-term peak admissions. Having raised four children in Douglas, he discounted the health issue altogether. Yuhnke argued on behalf of EDF members that they had a right to clean air and that the current SO_2 standards of 0.5 ppm over three hours did not protect the asthmatics among them. In fact, one of every six hours in this area had some time when the 0.5 ppm level was exceeded for six minutes or longer. Thus it was "impossible to have a daily routine without being exposed to SO_2 levels shown in medical research to have significant effects on asthmatics." He then presented an EDF video showing in a speeded up three-hour time sequence the actual motion of the plume as it hugged the ground, which segued into June Hewitt's filmed discussion of her problems with these short-term fumigations.[63] Cause and effect were deftly made, and a loud cheer went up.

Videotaped testimony at the Cochise College hearing provides a rich sample of community attitudes in addition to expert testimony. Among those present was former governor Jack Williams, who described himself as "an old man speaking in favor of an old smelter" that deserved an eighteen-month reprieve. But the tune had finally played out. In hindsight, the argument that stung the most during the years leading to shutdown was Yuhnke's linkage of the Arizona and Sonora smelters with acidic deposition in the Rocky Mountains. The health impairment of asthmatics, while deftly argued, was probably less important.[64]

That the company's credibility on Douglas was drawn into question owed much to the efforts of Richard Kamp, the third member of the team. If Robinson's strong suit was in networking and Yuhnke's was in science-based litigation, Kamp's was in generating and using information. Bright, personable, persistent, Kamp soon became a major player, and his role bridges events covered both in this chapter and the next. In subsequent years he became highly knowledgeable on border pollution issues and most recently has served on the Expert Advisory Panel for the NAFTA

Environmental Commission's pathbreaking study on the long-range transport of air pollution in North America. But he got his start in smelters.

Arriving from Chicago in 1974, Kamp was one of the many earnest seekers for a simpler way of life who found their way to southern Arizona. He settled in Naco, a small town on the Sonora border with huge vistas capturing spectacular desert sunsets. Not having grown up with the Douglas plume, Kamp was shocked by its frequent visitations from over forty miles away and wondered about the effects on his family's health. He could see the plume every day, including pollution from the Mexican smelter at Cananea. Soon he was picking up signals of economic distress from the farmers who, he learned, had been burned for years but could do little about it under the smoke easement regime and the small damage settlements PD was willing to pay. This personal reaction is what first set him off on the quest to learn about smelters and especially about the DRW. What he brought to bear on the smelter issue was a disciplined intelligence and a social conscience, and he soon became an expert on the border environment.[65]

Having contacted Priscilla Robinson at SES in 1989, Kamp received a small grant from her to study visibility issues and began monitoring the plume in time-sequence photography from the air and on the ground. Living with this daily irritant on the Naco plain, he became adept at gauging the drift and amount of pollution by looking at the smoke.

While learning about smelter technology, Kamp befriended some PD employees who clued him in about the actual behavior of the plume, how it drifted on diurnal currents into Mexico at night and sometimes blew back with a rush at dawn. When winds blew north, as they often did in the mornings, the converters (which produced the most SO_2) were cut back in order not to trigger the company's monitoring stations set at head level in an arc within five miles of the plant. These converters could be "batched," matte copper being quickly added or removed. But the roaster and reverberating furnaces, producing less SO_2 but more particulates, were costly to shut down. The winds tended to blow north by midmorning, and then this situation normally terminated by early afternoon, when winds and/or rising temperatures raised and dispersed the smoke over a broader area. By late afternoon, winds shifted to the south. That meant that "relative smelting freedom occurs until the following morning," and the smelter could be run full tilt into Mexico where there were no monitors.

Kamp's information was substantially correct. By PD's own account the March 1980 sulfur balance showed 49 percent sulfur coming from the roaster/reverb stack; 40.4 percent coming from the converter stack; and fugitive emissions from all operations accounting for the remaining 10.6 percent. A study confirmed that 70 percent of the total concentration for

1972 was measured during the six-hour period between 11:00 A.M. and 5:00 P.M. at Douglas. Indeed, it had been known for some time that "the high afternoon concentrations often exceed the 3 hour ambient air quality standard established to preserve health and welfare."[66]

Kamp also learned that there were no sulfur controls on the plant, and what emission control there was came almost entirely through a dispersion and production curtailment system or SCS. Engineers had "learned the ins and outs of SCS well enough to avoid excess violations," he wrote, but actual emissions were not being truly monitored by the incomplete network of ground stations.[67] Kamp passed this information back to Robinson, who soon realized that he now knew more than she did about how smelters worked. That it was profitable to run the antiquated Douglas works became clear to Kamp. In her testimony before Senator Stafford's committee that summer and then before the Waxman subcommittee in December 1981, Robinson used his data and an EDF analysis on the economics of smelting to show that the polluter had a competitive advantage over controlled smelters. By 1982, Kamp was operating with Robinson and Yuhnke as a team.

Meanwhile, the Cochise Smelter Study Group had been founded in Bisbee by Kamp and Michael Gregory in 1979 to study health and visibility issues. Members of the Bisbee group owed nothing to Phelps Dodge, which had closed down its Bisbee operations in the mid-1970s; they were very concerned with the Douglas plume, which reached as far as Tombstone to the north. But their efforts to raise the visibility issue under the rubric of the Clean Air Act were ignored. The idea was to cite Douglas for impairing visibility in nearby national monuments. Kamp's visuals showing what he called an "Arizona Highways of the Air" did not catch on. Their comments on the probable health effects of smelter smoke on asthmatics fell on deaf ears. In February 1982 Kamp and Gregory drafted a well-written pamphlet entitled *What's in the Smoke? A Breathers' Guide to Douglas Smelter Pollution*, but this, too, failed to elicit action from the congressional delegation or the state. In 1982 Greenpeace offered to help Robinson with the smelters: they pulled off the smokestack caper at San Manuel and said they would help Kamp's group with the media. But otherwise the Smelter Study Group's work on Douglas was being ignored by everyone, which was discouraging, and their funding ran out. Kamp eked out a living as a mechanic and Gregory as a printer. In spring 1983 the group shut down.

Almost immediately they hit pay dirt with the Mexican issue, and the group came back to life as the Smelter Crisis Education Project. It had been known for some time that Mexican interests were planning a large new smelter at Nacozari, which might or might not be controlled. This proposed project had already been reported in the *Breathers' Guide*, which also made the point that the Mexicans would be unlikely to control their old plant at Cananea or the new flash furnace at Nacozari as long as Douglas polluted freely into Mexico. But this wasn't news until concerns over World Bank financing for the old Cananea smelter broke that next spring.

Having heard that the World Bank through the International Finance Corporation was loaning money to expand the uncontrolled smelter at Cananea, Congressman Jim McNulty and Senator Dennis DeConcini, both Democrats with a labor constituency, became alarmed over what this financing portended for the depressed Arizona copper industry. In May a hearing was held in Tucson on this proposed loan and on the larger subject of loans to foreign companies in competition with U.S. industry. Robinson, who spent a great deal of time on the hearing and worked with McNulty to make the hearings go well, recalled that "he was very appreciative. The hearing was a big love-in, of unions, environmentalists, industry, etc., just before the strike" in July. Environmentalists testifying were Gaye Page of the Arizona Clean Air Coalition, Mike Gregory of the Smelter Study Group, and Robinson, followed by a joint press conference.[68]

While preparing the *Breathers' Guide*, Kamp realized that nobody was linking Douglas to the Sonora smelters, which were already being seen as a foreign threat. Then McNulty's hearing showing that American money through the IFC was going to "competitive" and dirty copper development on the other side of the border gave him and Gregory a platform to internationalize Douglas. At last they had the winning issue. "An international air pollution crisis of unprecedented magnitude is developing along the Sonora-Arizona border," he wrote in a request for project funding in September. "I have been immersed in smelter pollution . . . for 5 years, and on the international issue for 6 months," but now the project needs full-time work, and "if there is no funding, I can't do it."[69]

Then a friend on the staff of *Pay Dirt*, the Bisbee mining journal, passed on the information that Nacozari, which was designed to accommodate pollution control technology, would probably go into production initially without an acid plant. This tipoff was confirmed when Kamp went to Cananea with some reporters and then visited Nacozari, where he was told there would be controls. However, when he and the chief Mexican engineer got drunk together he learned that this statement was a lie: there would be no acid plant and pollution controls. The *Arizona Daily Star*'s Jane Kay was also on this trip, and Kamp urged her to do an exposé, saying if she did not do the article, he would. "She winced, but in nine days came up with a superb, hard-hitting report."[70]

The Mexican linkage took off, giving tremendous leverage to the team. In June, Priscilla Robinson reported to her board that the governor's office was considering a suggestion to take the lead in opening talks with Mexico about pollution control. As well, the attention given to EDF's work on acid rain was greatly enhanced by playing the Mexican threat card. And Kamp and Gregory's group was reborn with the Mexican smelter issue.

Around then the term "Smelter Triangle" (more usually called the "Gray Triangle") was coined by Michael Gregory to both locate and visualize the coming threat of transborder pollution from Cananea, Nacozari, and "Old Smokey." Kay used it in her August 28 article, giving credit to the Smelter Study Group. Kamp first used it in a long newspaper column published on October 23. "It appears," he wrote, "that by mid-1985 there will be more sulfur coming out of this smelter 'triangle' than for any equivalent square mileage on the North American continent: more than 3,000 tons per day of sulfur dioxide."[71] Without remedial action, southeast Arizona would become a pollution hellhole. From then on until shutdown, maps of the Gray Triangle became a fixture in Arizona and national press reports. In Colorado, Senator Gary Hart and Representative Timothy Wirth took notice of the threat to mountain lakes, and Sandy Graham's articles critical of the air impacts of copper smelters in the mountains began appearing in the *Rocky Mountain News*.

In October Robert Wick, who with his brother Walter ran a chain of newspapers from Sierra Vista, hired Kamp to report on smelter issues and gave the Smelter Crisis Project enough funding to get started. Wick, a Republican from an old Ohio family who was also a sculptor, had become fed up with the "public be damned" attitude of Phelps Dodge. Receiving a check every two weeks was a refreshingly new experience for Dick Kamp. In his new, combined role as reporter and as citizen activist, the personable and persistent Kamp became highly effective and disarming.[72] Using Robinson's mailing lists, Kamp and Gregory formed Groups Against Smelter Pollution (i.e., the second GASP), a Bisbee grassroots organization, to mobilize Cochise County. The radical feminists running GASP soon clashed with Yuhnke, being impatient with his deliberate step-by-step approach. Kamp and Robinson had second thoughts, but under new leadership GASP persisted as an activist pressure group until the smelter shut down. In 1985 the smelter project became part of the newly formed Border Ecology Project, a broader vehicle for attracting outside funding that continues today.

Internationalization worked brilliantly for Kamp, who was by now involved in several cross-border issues, including the transport of hazardous materials, health, and the effect of smelter smoke on Mexican small farmers. More on the Mexican phase of his activities will be discussed in the next chapter. Suffice it to say here in conclusion that the other strands all mattered: the Clean Air Act that ended the smelters' isolation and empowered EPA regulators and citizen activists alike; PD's

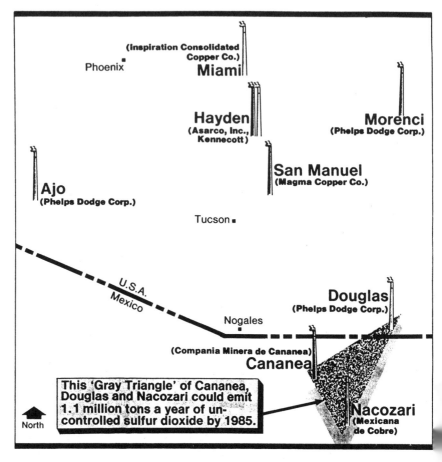

Map of the Gray Triangle. (Judy Margolis, *Arizona Daily Star*, 1985, p. 13e; courtesty of the *Star*)

tough-minded and successful survival strategy in the face of growing public hostility and the escalating costs of fines for air pollution violations; the strike; the acid rain issue and (secondarily I think) the health issue; and the changing social base of southeast Arizona. But how important was the Gray Triangle? The cross-border linkage with Mexico was indeed significant—and had been all along, especially since an important outcome of the Trail dispute was to enable the smelters to continue polluting virtually at will into Mexico. This issue is the topic of the next chapter.

CHAPTER SIX

The Gray Triangle

When James Douglas, the Canadian-born mining engineer and metallurgist, founded the smelter town named after him in 1900 and for many of the intervening years until 1987, when the plant was shut down, smelting at Douglas had been a transborder operation. It was well located to process ores from Bisbee and Nacozari, some sixty miles to the south by rail in Sonora, where a mine and mill were operated by Phelps Dodge from 1895 until 1960. (In fact, one of the last contracts Douglas had was to process ore from La Caridad mine, now owned by Mexicana de Cobre, whose modern smelter at Nacozari came on line in 1986.) But with the closing of these mines the DRW no longer had a place in Phelps Dodge's long-range strategic planning. Being poorly located to serve the main producing mines at Tyrone and Morenci, it was never intended that the smelter be rebuilt.

That the DRW had been designed to process concentrates from Pilares as well as Bisbee meant that it was a truly transnational operation for most of its long life. And when at last (in the early 1980s) the long-anticipated linkage with transborder pollution on the Mexican border was finally made, Douglas's fate was sealed. Still, the corporation was taken by surprise when the Gray Triangle emerged as a public policy issue at the national and local levels.

Corporate interest in Mexico never disappeared entirely. Along with Anaconda, PD retained a minority stake in the Pilares mine at Nacozari until 1994. In the early years Phelps Dodge had owned a mine at Cananea, which was never a large producer, and in the late 1930s ran the San Carlos lead mine in Chihuahua. However, even when it operated the underground mine at Pilares de Nacozari, PD holdings in Mexico were always dwarfed in size and importance by the Mexican assets of ASARCO. Concentrates from La Caridad at Nacozari were shipped to Douglas and the ASARCO smelter in El Paso.[1] To this day ASARCO holds a one-third interest in Mexicana de Cobre and the new Nacozari smelter.

With its twin stacks and the plume, DRW was both an economic and a social fixture on the border. The history of this transborder zone is intimately linked to Douglas, which was for many years a classic company town—hierarchical, segregated, but with a strong sense of place. However, by the 1980s it was rapidly becoming a Hispanic border town in contrast to Bisbee, the Anglo town. Management's move to Phoenix accelerated the transition from company town to border city, but many old-timers remained. For them, life without the smelter (and the plume) was inconceivable, even though the socioeconomic base of Douglas and Agua Prieta was changing rapidly. When it came, several hundred good jobs were lost, but the DRW's closing did not disrupt the diversifying economy of this coupled nucleus for long. Local merchants were glad to forgo the competition that Phelps Dodge Mercantile had given them for the smelter worker's dollar. Even so, opinion in the town of Douglas was deeply split over the shutdown issue.

For years, farmers on both sides of the border had received payments for smoke damage to their crops. And if payments were small, the fact that Mexican nationals had been included since 1937 and during the war years when the smelter reached peak production showed the importance of the smelter's economic reach into the countryside on both sides of the border. As was typical when smelters operated in rural areas, the pollution came with the benefit of an urban market for the farmers' crops. As well, since the late 1940s, Hispanics had access to well-paying jobs at the smelter. In sum, for many on both sides of the border, that plume still was "the smell of money and jobs."

As already discussed, local protests against pollution were ineffective until outside forces opened new avenues for action and new ways of thinking in the hitherto isolated towns of southeastern Arizona. The roles of the EPA and the environmentalists and the corporation's response are by now familiar to readers of this book. These forces swept the locals along: previously outgunned American farmers received higher payments; Mexican farmers on ejidos and small ranchers, feeling *agachado* (inferior) vis-à-vis the gringos, would not have dared to speak up against Phelps Dodge. Now they too were questioning the corporation's smoke damage decisions and were willing to be interviewed by reporters.[2] However, one should not press the point.

Given the deep, long-standing social and economic ties in the border region, it is very significant that pollution was not seen as an international issue until very late in the game. The plume was simply a fact of border life. Aside from the 1937 court case, the Mexican government was not making an issue of it. (The Mexican government did become active in the 1970s when residents of Juárez protested to the International Boundary and Water Commission about pollution from ASARCO's lead and copper smelter in El Paso.) And when the plume from Cananea drifted north over the years, it was sufficiently diluted by the time it crossed into U.S. territory that the American authorities were not making an issue of air pollution from the Mexican plant either.

But in 1983 this old fact of border life was reframed and rapidly became transnationalized, thanks to the McNulty hearing and to mounting concerns about the huge increases in pollution expected from the new smelter at Nacozari. By the time a town meeting was called in Bisbee for June 6, 1985, to protest transborder pollution, a new public discourse was being articulated just as much in Douglas, Bisbee, and Tucson as it was in the halls of Congress. "If you live in Southeast Arizona, Northeast Sonora or Southwest New Mexico," the Bisbee circular proclaimed, "no matter who you are you will be affected. When the 930-foot Mexicana de Cobre Nacozari Smelter stack fires up in 1985 and joins the Douglas Smelter 55 miles north, and the Cananea smelter, we will live amidst 2,700 tons/day of sulfur dioxide and particulates." A "Gray Triangle of Pollution" was in the making. These transborder linkages were symbolized by the presence of Dr. Alberto Durazo, a pediatrician at Cananea, on the speakers' platform. (Later, he and his wife formed Projecto Fronterizo de Educación Ambiental, one of the leading border NGOs.) Joining him on the platform were Dr. John Abbott, a Bisbee physician who talked on health effects, Linda Lewis from Congressman McNulty's office, and Dick Kamp. Mexico, health, Congress, and grassroots activism were thus conjoined. How had this bundling of issues come about?

Certainly one factor was the belated but forthright stand of Arizona health and environment officials who were worried about the health effects of smelter pollution. Prodded by the EPA and citizens' groups to be a more active enforcer of air quality standards for Arizona smelters, and responsive to Governor Babbitt's anti-Douglas stance, it was now conceivable for the first time that the state might not renew the Douglas operating permit. Dave Chelgren, director of the Bureau of Air Quality Control, had for years avoided confrontation with the smelters. But now he prepared a report on the two old plants plus Nacozari for Congressman McNulty's border pollution meeting, which was held in Douglas in July. Even with Douglas shut down, pollution loadings would more than double when the two uncontrolled Mexican plants in the triangle came on line, Chelgren said. These plants were so close to each other that a degradation of visibility could be expected. "Also, experience with the Douglas smelter over the past several years has indicated vegetation damage from large SO_2 emissions. These effects may certainly be experienced on both sides of the international border, as the Mexican smelters emit more SO_2," he concluded.[3]

As for Douglas, the last full year of production was in 1981, when it emitted 910 tons of SO_2 a day. Recent emissions had been less than half that figure, Chelgren said, and in any case the smelter would be forced to close under the Clean Air Act on December 31, 1987. Cananea emitted 314 tons a day in 1982, and if, as planned, the facility was upgraded with an oxygen-enriched furnace, they could be emitting as much as 1,000 tons a day through a 900-foot stack. Nacozari, utilizing an Outukumpu flash furnace, would be emitting 1,200 tons a day through a 932-foot stack. With winds blowing from a southerly direction, a considerable portion of the expected emissions from the Mexican smelters would be transported into Arizona. Neither of the Mexican smelters planned to have an acid plant, and without controls "violations of EPA's ambient air quality standards may be expected to occur." As well, Chelgren's statement showed that the Mexican linkage was by now fully established at the state government level.

For its part, the Arizona Department of Health Services (ADHS) had been concerned before about the declining water quality of the San Pedro River, which flowed north into Arizona from headwaters near the Cananea mine and smelter complex in Sonora. Copper tailings escaping from settlement ponds at Cananea impacted downstream agriculture and wildlife in Arizona in 1977–1979 and again in 1989. Following discussions with Compania Minera de Cananea and the Export-Import Bank in September 1981, management agreed to relocate the smelter's tailing ponds in a different watershed, so that any escape would not flow north into Arizona.[4] Groundwater depletion from overdrilling on the Mexican and U.S. sides was also an issue and remains so today. In short, the state had an interest in resolving transborder pollution water issues with Mexico well before the Gray Triangle emerged and with it the air quality and health issues associated with smelters.

For his part, Governor Babbitt in his poststrike posture was genuinely concerned with the social equity issues raised by pollution from Douglas. That an Arizona smelter had for years been damaging crops in Mexico struck him as unfair, and by 1985 he was saying so in public. The smelter, he said, "can discharge onto our Mexican neighbors amounts of sulfur dioxide which American citizens would never be forced to endure."[5] He called on the federal government to act immediately to protect Mexican citizens from hazards created by the Douglas smelter.

Well before this, Babbitt had taken an active interest in transborder issues. In fact, he was one of a new generation of governors who saw the necessity for cooperation with their Mexican counterparts. He became active in resolving disputes over the importation of Mexican produce, having headed off an attempt by the Texas-based American Agricultural Movement to close the Arizona border during his first week as governor. "This made a deep impression on me," Babbitt recalled, "and it was an experience that shaped all my dealings with Mexico." Then he helped to resolve the famous tomato war when Florida growers tried to block Mexican exports. During the energy crisis he tried to get ARCO to distribute Mexican refined products in Arizona and discussed the possibility of entering into a joint venture with the state of Sonora to refine oil supplied by pipeline from Guaymas. At a time of difficult relations between the two federal governments, he discovered that "an enormous vacuum existed" in the transborder relationship. Babbitt, who spoke Spanish from his Peace Corps days in Peru, found that the Mexicans were astonished that anyone would know their side. In 1985, "I met with [President] de la Madrid and his cabinet on how to approach the Americans," who rely so heavily on lobbying. "The immigration issue was really getting them, for it wounded their strong sense of sovereignty and . . . [they knew they could not avoid] an issue that was going to affect Mexican employment and economy." He also helped on the drug issue, "which pleased de la Madrid a lot." It is in this context of engagement with Mexicans on other issues that his conversations with them about how to deal with *contaminación* (pollution) in the Gray Triangle took place.

Before discussing Babbitt's meetings with the Mexicans, it is well to summarize why the Americans were so concerned. Recall that the Mexican linkage first won public acceptance at Jim McNulty's Tucson hearing in July 1983, which focused attention on the Inter-American Development Bank loans to expand Cananea and build new capacity at Nacozari—all in competition with the depressed Arizona copper industry and without requirements to install controls. The plans for Nacozari did include an acid plant, but there were doubts about when, if ever, it would be built. After this hearing, the issues of Mexican competition and pollution were now yoked together in the public's mind.

There was a subtext in this approach, which was the demonization of Mexico to protect American economic interests. As well, environmentalists used the specter of Nacozari to engage the public, especially in states with high mountain lakes, such as Colorado, New Mexico, and Wyoming. But the core participants in this debate also framed the issue as one of reciprocity. In Babbitt's words, "Unless we act promptly [to protect Mexicans from Douglas], we can hardly expect a reciprocal response from the Mexicans with respect to Nacozari and Cananea."[6]

As Arizona's congressional delegation and Governor Babbitt became engaged, ADHS began to stiffen its requirements for smelter operating permits and to deal seriously with the smelter health issue, although the agency initially was still wary of getting in too close to a Democratic governor at odds with a Republican-dominated and pro-smelter legislature. Babbitt was by now accessible to Priscilla Robinson, especially after he had been forced to call in the National Guard to keep public order during the Morenci strike. While Robinson worked with ADHS and the governor, Dick Kamp fed information to McNulty. With their strong union ties, both McNulty and DeConcini became committed to resolving the Gray Triangle issue. Mexico could not be asked to put on controls with Douglas still operating as it was. In Robinson's words, "the importance of the Mexican affair was that once committed to pressuring Mexico to put an acid plant on Nacozari, McNulty and DeConcini were committed to closing

Douglas, which had been polluting Mexico for decades. Once they made this commitment in the press, there was no backing out." By 1985, she recalled, "the chief objective of the environmentalists was the PD operating permit, the Mexican connection, and linking the two."[7]

For the NGOs working to shut down Douglas, the Mexican pollution threat was made to order. EDF used it to highlight the acid rain issue, as has been discussed. In Congress, an unlikely coalition of national environmental groups and the copper industry seeking protection against foreign copper imports was mobilized to oppose the loans to Cananea and Nacozari. This group introduced highly restrictive environmental and protectionist legislation that would have penalized Mexican copper companies producing without environmental controls. (Being closer to the scene, the Arizona environmentalists had misgivings about this approach.) As well, certain Democrats from the West, who had been fighting against attempts to weaken the Clean Air Act, now focused on the Mexican smelters linkage. Senator Gary Hart and Representative Timothy Wirth from Colorado were particularly concerned about the effects of high mountain acid rain from smelters. They coordinated their legislative strategy with EDF's Bob Yuhnke, who because of the acid rain issue had access to Administrator Ruckelshaus to discuss the Mexican linkage.

To backtrack still further, air quality issues along the border had arisen in the late 1970s, particularly automobile and industrial pollution in the twin cities of Juárez–El Paso and Tijuana–San Diego. Having been raised by state, local, and civic activists, the profile of this issue was greatly enhanced when the EPA declared these border conurbations as out of compliance with the Clean Air Act. The copper issue played into this dynamic as well, enabling environmentalists and western congressmen to put sustained pressure on the EPA and the State Department to explore avenues for obtaining an air quality agreement with Mexico. Writing soon after the event, political scientist Steve Mumme provided a good perspective on how it was done:

Led by Arizona's governor and congressional delegation, borderstate advocates of an agreement on smelter pollution formed a working alliance with California's congressional delegation, which was then seeking priority action on binational sewage pollution in the San Diego–Tijuana region. According to Mumme, "the Arizona and California delegations were supported by New Mexico and assorted congressmen interested in the emerging acid rain problem in the Northeast. Together this coalition was successful in placing border environmental problems as a set on the August 1983 summit between President Reagan and Mexican President de la Madrid in La Paz, Mexico."[8]

As a result of the La Paz summit, the Border Environmental Cooperation Agreement of 1983 became the first comprehensive transboundary pollution agreement between the two nations. And in 1987 the Gray Triangle was explicitly recognized in Annex IV of this executive agreement. The upshot was that the United States was engaging with Mexico about air quality issues several years before it was willing to seriously engage the Canadians over acid rain, which was Canada's priority.

This accomplishment in the area of environmental relations is remarkable, given the rising tensions in other areas such as finance, migration, and the Central American wars. In view of the close working relationships that exist today between officials at all levels, it is hard to imagine that just a decade ago in these pre-NAFTA years there was no climate of cooperation. From the Mexican perspective, they were starting from a long history of little contact between the two diplomatic corps. At that time, "few [Americans] knew how to get along with us and we in the U.S," Alberto Székely, the lead Mexican negotiator, recalled. At the time, "Reagan was hostile, and couldn't care less about Mexico. John Gavin [the American ambassador], who had such good Mexican family roots, could hardly get Reagan's ear on anything," and the Mexican press, especially the left, was hostile to them both. Such were the unpromising beginnings as seen from the U.S. side, recounted by Cliff Metzger, a foreign service officer working on La Paz who was stationed at the embassy at that time.[9]

Because the two nations were becoming ever more interdependent, necessity was the mother of invention. That it occurred first in the area of environment was "at least partly a function of increasing state [government] activism in environmental affairs," Steve Mumme observed.[10] In a number of well-crafted papers on border issues, he continued to follow environmental activity at the subfederal level. However, with the perspective that a decade brings on these events, more is now known about the play of interests. Why did Mexico sign on to what would surely be a costly agenda of environmental remediation? And what of chance and contingency, those old friends of historians and the bane of political scientists?

State and local concerns over water issues and toxic spills were more important than air in bringing the two sides together. As Metzger recalls, the United States had wanted an agreement on toxics in 1992, and it was the Mexicans who first suggested a cross-border agreement. He and Székely negotiated it for six months, but nothing was happening. Then, with the Californians hollering about Tijuana sewage and the Mexicans getting impatient, Reagan asked what he could sign when he got to Mexico for the upcoming meeting with de la Madrid in La Paz. "We hurriedly redrafted the accord and he signed it with no idea of the content." The result was that "we got a lot done under La Paz, lots of mileage." The La Paz working groups, meeting every six months, created a new relationship. It was "the blind luck of Ronald Reagan."[11] Historians puzzle over this president: was his worldview simplistic or just simple to understand? Reagan's performance at La Paz provides at least part of the answer.

Not that de la Madrid himself saw La Paz as an environmental meeting. Indeed, in the chronicle of his presidency for 1983, the agreement is mentioned only briefly, and it is clear that two themes dominated this meeting: bilateral economic relations and Central America.[12] For their part, the Mexicans felt the pressure from San Diego to do something about sewage. "That was why [de la Madrid] signed," Székely recalled. San Diego's congressman attended the La Paz meeting. At that time, the White House seemed beholden to Reagan's Southern California cronies, and the Mexicans were very aware of and concerned about the weight of California politics, about which they were poorly informed, however, because of a weak consular staff in California at the time.

Drafting on the Mexican side was done by Alberto Székely, legal adviser at the Foreign Ministry, and Roberta Lajous, director of the North American division of SRE and wife of the foreign minister. They worked together on border matters for six years, in contrast to the Americans who changed personnel frequently.[13] The American position was drafted by the State Department's Metzger.

The agreement signed in La Paz committed the two countries to address border environmental problems within a comprehensive framework. The EPA and Mexico's newly created environmental ministry, the Secretaría de Desarollo Urbana y Ecología (SEDUE), were designated as national coordinators for each country. The agreement established a series of Annexes to address specific pollution problems including Annex I, which deals with the Tijuana River sewage problem. Over the years, the scope of La Paz was broadened to include hazardous materials emergency response (Annex II), the handling and transport of transboundary hazardous waste (Annex III), emissions from copper smelters (Annex IV), and urban air pollution (Annex V). A series of bilateral working groups were also established, both to examine the issues and to ensure compliance with the objectives of the Annexes. There was no formal enforcement mechanism.

The Gray Triangle had been established as a U.S. issue well before a formal Annex IV on smelter pollution was signed on January 29, 1987. The fact that it was an issue for the United States much more than for Mexico goes a long way toward explaining the lag of three and a half years, which is an eternity in politics. In fact, on economic grounds alone there was no reason for Mexico to agree to an expensive acid plant or to put controls on Cananea. Paying for expensive equipment to produce by-products (acid and gypsum) for which there was no real market advantage was not the optimum business plan.[14] As well, Mexico's economic crisis since 1983 had sharply reduced the funding for investment in environmental programs, placing expensive regulatory projects on the back burner.[15] Basically, where the Americans saw a Gray Triangle producing a "pollution hellhole," the Mexicans saw only a thinly populated desert region remote from Mexico City. Furthermore, the Americans had been polluting into Mexico for years, so why should the Mexicans worry about sending smoke the other way? These arguments made sense, but they resonated more with the internal logic of the Americans, as they stepped up the pressure on Nacozari, than with the Mexicans looking for a way to respond.

The best weapon that the environmentalists and their allies had in their arsenal of arguments for shutting Douglas down was the Mexican threat. Framing the argument in moral terms as an issue of reciprocity also played well in the United States, and it probably made the Americans feel better as they pressured Mexico to act on something that seemed certain to stir a nationalist reaction.

Furthermore, the Americans entered these negotiations with the credibility of having already put controls on ASARCO's copper-lead-zinc plant in El Paso. This longtime polluter had been worrying local health authorities for years, and in 1972 an investigation conducted under the auspices of the National Centers for Disease Control found that children living near the plant in Smeltertown had high lead levels in their blood. The IQ scores of El Paso children fell with the incidence of exposure to lead. Health authorities in Ciudad Juárez just across the river found a similar distribution of lead levels in children according to their distance from the El Paso smelter. Lead emissions were significantly abated by 1974, but emissions were still beyond EPA standards. Mexican complaints about continuing pollution reached the IBWC, the binational agency with jurisdiction for boundaries and water. The EPA then brought pressure on the state of Texas, which had lower pollution control standards than Arizona, to enforce the national standards. ASARCO responded by installing SCS, which was both expensive to operate and environmentally unsustainable. In 1982 it was required to install a double absorption acid plant to treat converter gases, and eventually the smelter was upgraded with a modern oxygen-enrichment system developed in Germany and perfected on-site.[16]

One of the two premier polluters on the U.S. side of the border was now controlled. "This gave the U.S. credibility when it came to pressuring Nacozari," Metzger recalled.[17] Douglas and the Gray Triangle were next.

For their part, the Mexicans readily accepted the fact that smelters on both sides were guilty of polluting. But as Babbitt recalled, "They were of two minds on this one. They were sympathetic to the pollution control issue but rejected badgering from the U.S. or any perception of badgering from the U.S." In talking with President de la Madrid about Nacozari, he raised the issue of *contaminación* (pollution) as a problem on both sides of the border. Out of conviction and in line with American thinking at the time, Babbitt framed it as a problem of parity. The Douglas works were "a rogue elephant that was my problem," he told the president, but it was something that they both faced together. "I told him Arizona would have to comply first before I could get aggressive on the Mexican issue." Realizing that de la Madrid did not have staff working on it at the time (1985), Babbitt did not press the point and "let it lie there this way at the time."[18] His position was consistent: "I have always believed that the Mexican government desires reciprocity in controlling smelter emissions," he wrote to Administrator Lee Thomas, "and that they have every right to do so."[19]

Probably the most difficult problem for Mexico was getting their position together to deal with what was seemingly not a good issue for them in the bilateral relationship. SEDUE, the newly created environmental ministry, was still being organized and had little influence within the Mexican government. Halfway through the negotiating process the first director was forced to resign in April 1985 over a nonrelated issue, and a large staff turnover ensued. Mexico also experienced the first of those drug flare-ups that periodically capture the undivided attention of the American press. Then the government came under intense criticism for its inept response to the massive earthquake that devastated Mexico City that summer. SEDUE improved under Manuel Camacho de Solis, the new director, who as a Princeton-trained economist also had a more international perspective. Given carte blanche by President de la Madrid to solve the smelter problem, Camacho moved to address American concerns about Nacozari and Cananea.

What the Americans did not know, or at best did not appreciate at the time, was the linkage of the smelters to Central American policy, over which the two capitals were feuding. Relations were tense. Mexico was extremely worried about Nicaragua, believing that the United States might send in troops at any time, which would vastly complicate their internal politics. Not knowing what to expect from the Reagan administration, Mexico was making policy on a day-to-day basis. From their perspective, Central America was the problem, not the border, and they did not want to be perceived in Washington as hostile to the United States on other aspects of the relationship. Thus they, too, wanted to move on the smelters.[20] Evidently, in the vacuum then existing in relations between the two countries, U.S. negotiators never appreciated that they were pushing on a locked door with a key in it.

In July 1985 SEDUE signed a letter of intent pledging not to expand capacity at Cananea without prior permanent controls and to have an acid plant in place at Nacozari by March 1988. (Plans to expand Cananea had already been shelved.) Nacozari would still open and then operate for almost eighteen months without controls. However, until the acid plant was ready, exceedences beyond the ample limits set by SEDUE would be prevented in the interim by interrupting production when data from the monitors and atmospheric conditions warranted it.[21] Basically the same system worked out for Trail in 1938 would do for Nacozari almost half a century later. This approach disappointed those on the American side who anticipated months of heavy pollution from Nacozari just as they were pushing Phelps Dodge to avoid any violations of the air standards until shutdown in 1988 (at that time, this date was still somewhat less than certain). However, in a joint communiqué, the two countries "agreed that the control of the Douglas and Nacozari smelters should be linked so that interim and permanent controls will be applied."[22]

A problem for Mexico was how to fund the capital for the environmental controls required for the two Mexican legs of the Gray Triangle, given that the Reagan administration was adamantly opposed to committing U.S. public loan funds to help Mexico, the polluter, solve its problems. The transborder financing of equipment by the Export-Import Bank, seen by McNulty, Babbitt, and the environmentalists as a positive incentive for Mexico to comply, was therefore not an option. The American government strongly resisted this approach, insisting instead that each country was responsible for covering the costs of its own pollution.[23]

Under "the polluter pays" principle, the Americans might have been liable for damage to Mexico, but as copolluters neither country had any interest in dusting off this classic but largely unworkable doctrine arising out of the Trail smelter dispute many years before. There was never a question of transborder restitution for damage to farmers or sovereignty on the Mexican side or the possibility of claims for anticipated damage to mountain lakes from the Nacozari plume on the U.S. side. The "polluter pays" principle would have been even more ill-suited to this situation, because international agencies like the World Bank and the IDB were making development loans to Mexico without mandating air pollution controls.

The United States exercised strong influence in these international lending institutions, and the Cananea project was canceled, which the Mexicans resented. Cananea, a parastatal enterprise, had symbolic value because of the industrial strike in 1907 that was a precursor to the Mexican Revolution. But it would have been difficult to put controls on that plant, which had the same reverberatory furnace technology as Douglas. The Cananea expansion was canceled as much for business reasons as from congressional pressure. Scrubbers eventually were put in, a minimal control, especially as production increased fourfold when the concentrator was expanded with old equipment purchased from Douglas.

Mexicana de Cobre, which was building the new Nacozari smelter, was controlled by Jorge Larrea, a Mexican engineer who got his start in copper by buying out ASARCO mining interests in Mexico. (The American company then took a minority interest in Nacozari.) An aggressive tycoon known for cutting corners in his business deals, Larrea wanted to buy out Nacional Financiera (NAFIN), the government development bank that partially funded the new plant and also guaranteed the loans from U.S., Japanese, and Canadian banks that financed the remainder of the facility at Nacozari. Knowing that he was sucking up to the government, SRE arranged for him to attend a small luncheon being given for Secretary of State George Shultz by Bernardo Sepulveda, the foreign minister, in Mexico City that July. Larrea was flattered to be asked. In return he accepted SRE's suggestion that he build the acid plant, for which NAFIN, the government partner, then secured a soft money loan.[24] As a result, 15 percent of the $382 million raised by NAFIN and the private sector to build the Nacozari works was sunk into pollution controls.

On June 1, 1986, Larrea signed an agreement with SEDUE and the Ministry of Mines that made the opening of the new smelter conditional on starting construction of the acid plant, installing monitors, and holding SO_2 emissions to a maximum of 0.13 ppm in any single twenty-four-hour period.[25] President Miguel de la Madrid laid the cornerstone. Construction of the acid plant itself was supervised by FENCO, a Canadian engineering company, which brought with it $50 million in funding from the Export-Import bank of Canada and helped to train the workforce. A subsidiary of Mexicana de Cobre did the actual construction work. (A temporary setback occurred when the private investors headed by Larrea could not make dollar-denominated payments on their 56 percent ownership share of the smelter and NAFIN stepped back in.)[26]

Historically, Larrea said at the smelter's inaugural in June, almost all of Mexico's ore had to be sent to the United States for processing. Here, finally, "was the parting of the waters": Mexican ore could be processed at home.[27] Mexicanizing the mine and smelting operations was the equivalent of Aldridge's success in Canadianizing the Trail operation for the Canadian Pacific Railway some eighty years before. With respect to copper, Sonora was no longer a peripheral part of operations in Arizona. Yet in providing technical assistance to this state-of-the-art plant, the industry once again demonstrated its traditional willingness to share technology. It was a classic win-win solution: Larrea was well on his way to controlling 6 percent of the world's copper industry, and the Americans were now closer to getting the Gray Triangle under control with a smelter annex to the La Paz agreement.

To be sure, from the business perspective pollution was hardly the main story; the development of significant new capacity in Mexico just as prices had sunk to Depression levels worried American producers. Mexican ore production had not exceeded 30,000 tons per year until after 1967, when the Sonora mines were gradually expanded as an import-substituting and job-creating project, reaching 67,100 tons by 1979. Ore production levels only became worrisome to Arizona companies in 1980, when output jumped to 153,743 tons. Production dipped a bit during a 1983 strike but by 1986 had reached almost 243,000 tons per year. Basically, ore grades and by-products at Cananea and Nacozari were very similar to those in southern Arizona, and the same was true for mining and milling methods. But the transborder region was no longer integrated under American control.[28]

Concurrently, the old Cananea works were expanded to smelt 48,000 tons per year in 1986, with Nacozari adding another 80,000 tons per year, and this new plant's capacity was soon to almost triple to 165,000 tons per year. Nacozari copper was shipped for refining to factories in El Paso, Baltimore, and Mexico City, and large quantities of sulfuric acid were made available for fertilizer production and for export. From the air quality perspective, it was the prospect of the additional 417,000 tons per year of SO_2 loading from Nacozari that commanded attention. But from the business perspective, it was the arrival of a new producer that concentrated the mind. As the *Tucson Citizen* commented, "Try convincing an unemployed copper miner [here] that the United States should send millions of dollars to Mexico so its copper industry can expand."[29] Funding for Cananea was temporarily set back, but U.S. banks were quite willing to finance La Caridad, which became the world's third largest open pit mine as well as a world-class smelter. The Export-Import Bank financed the purchase of U.S. equipment, and as has been discussed, the old converter plant from Douglas was incorporated into Cananea. Meanwhile, to capture market share in the face of falling international commodity prices, Chile expanded the installed capacity of its nationalized copper industry. During the 1980s, its capacity to smelt concentrates increased from approximately 1.9 million tons to 3.4 million tons a year. None of this new capacity was controlled, which gave Chile a competitive advantage. (Later, remediation in the form of new technologies was applied in the 1990s.) And as the price of copper dropped to a low of seventy cents a pound in May 1987, the U.S. copper industry was challened as never before in the postwar era.

In sum, the La Paz agreement was part of a larger arrangement between Mexico and the United States as well as major changes in the international industry. A new and mostly modern copper industry was installed in Sonora over the objections of U.S. producers, but this new capacity would be subject to constant air pollution controls, a measure to which the two governments, each for their own reasons, supported.

Environmentalists following this issue were disappointed that La Paz did not go beyond consultations but pushed anyway for a smelter Annex. Unlike the IBWC, which had high-profile sponsorship by the two foreign ministries and regulatory authority to deal with border water issues, the La Paz agreement responded to the EPA, with its national agenda, and to SEDUE, a new, inexperienced, even chaotic agency. What most concerned the Mexicans was the off-loading of hazardous industrial and medical wastes into their country. They noted that under an existing hazardous wastes antidumping agreement, five maquiladoras on the border had appeared before U.S. courts in 1986 for dumping into Mexico.[30] Not until Annex III, governing hazardous wastes, was signed on November 12, 1986, were the Mexicans prepared to sign on to Annex IV, dealing with smelters in the Gray Triangle, which they did on January 29, 1987. However, their use of this linkage was fortuitous, and in fact the Mexicans were quite willing to accommodate American pressure on the smelters.[31]

Another factor in the delayed resolution of the Gray Triangle problem was division on the American side. Fitzhugh Green, the EPA deputy administrator for international affairs and a Ruckelshaus crony, was reluctant to engage the issue. At the outset there was no badgering of Mexico by the EPA, which claimed it had no jurisdiction over the Gray Triangle. Green told Babbitt that he too had no jurisdiction and should not get involved in foreign policy. By talking with de la Madrid and others about this issue he risked putting himself in violation of the Logan Act. "This was hilarious," Babbitt recalled, not at all intimidated by the prospect of falling afoul of an old statute nobody had ever heard of. Babbitt raised the smelter issue with the EPA two or three times and had a good relationship with Lee Thomas, the new administrator taking over after Ruckelshaus, but had the impression that Thomas was not being aggressive enough on this issue or being directly responsive to him. For their part, the environmentalists felt that the EPA was, in Dick Kamp's words, "caught in the regulatory process and under multiple pressures." Green seemed out of his depth, not in charge.

At the first meeting of the national coordinators in Tijuana in March 1984, the two countries agreed to establish working groups to address air, water, and the handling of hazardous materials along the border. San Diego–Tijuana sewage and the Gray Triangle were designated as priorities for binational action. But it required congressional legislation in a bill sponsored by McNulty before the EPA got involved in studying the transborder aspects of the smelter issue. The actual negotiating was done by David Howekamp, from the Air Division of Region 9 in San Francisco; Enrique Acosti was his counterpart from SEDUE.[32]

State Department officials were more forthcoming. Eager to establish good relations with Mexico, they took a forthright stand on the Gray Triangle at the Albuquerque hearings in 1985 and along with the governors of Arizona and New Mexico asked the EPA to put the screws on Douglas. For his part, Alberto Székely formed a very close working relationship with State's Mike Kozack on the smelter and hazardous waste issues. Officers at the embassy began sending information to Dick Kamp, detailing Mexican positions. In turn, Kamp fed information about the American positions back to the Mexicans at SEDUE.

Citizen diplomacy came into its own on the Gray Triangle smelter issue. The door to public participation had been opened by the Clean Air Act of 1970, but in his crossborder networking Kamp was charting new ground. For one thing, he monitored all the meetings of the air working group so that he was well informed on the status of the negotiations. For another, he spent a good deal of time with Cliff Metzger, the man at State assigned to smelters, and developed close ties to Mexican officials at SEDUE. A brilliant networker, Kamp was persistent and had great contacts. Working from his list of Mexico City home phone numbers, he got the inside dope from SEDUE officials, which he then used in talking with Judith Ayres, the regional administrator of Region 9 and Dave Howekamp. As Yuhnke recalls, "the Mexicans thought he was on their side, EPA people saw him on their side. Dick's role was to be the gadfly and a pollinating bee,"[33] busily networking among agencies and individuals who were themselves learning the habits of cooperation.

His opening to the Mexicans occurred in July 1984 when SEDUE officials visited Bisbee to assess the acid rain issue being highlighted by the Gray Triangle. Kamp went drinking with them in Agua Prieta, "and after an incredible number of beers Alfonso Oñate of SEDUE gave me my first good briefing in Mexico." At the invitation of SEDUE and SRE he visited Mexico City in October. Rosalda Gallendo from Governación (the president's staff) was sympathetic and got him into networks, playing this role until 1987 when she died in a car accident. Nidia Marin of *Excelsior* publicized the air issue on page one, which gave the Kamp visit "an incredible entry into official Mexico" when her story on the threat of pollution to the border also appeared on page one. In the Federal District, among other people he met was Fernando Ortiz Monasterio, the leading Mexican expert on air pollution who taught at the Colégio de México. Thanks to the publicity generated by this trip, Kamp had good access to the science attaché at the U.S. embassy, who sent him information on the Mexican position, and to Metzger back in Washington, who knew Mexico from his tour there and liked the smelter issue a lot.

Later, at a smelter conference[34] sponsored by the University of Arizona in February 1986, Kamp renewed his ties with Monasterio, who said that what the Mexicans were really interested in was the hazardous waste issue, whereas smelters were an American issue, which would account in part for the delays in bringing Annex IV to closure. He also met Roberto Sánchez, an economist at Colégio de la Frontera Norte in Tijuana, who became his closest associate on the Mexican side. (In 1994, Sánchez joined the staff of the Commission for Environmental Cooperation in Montreal and now teaches at the University of California's Santa Cruz campus.) Alberto Durazo, the pediatrician at Cananea, was a key informant as Kamp became knowledgeable about the Mexican smelters, befriended their managers, and developed a deep affection for the country.

When Annex IV was finally signed in 1987, it bore the stamp of this NGO activity in the space and importance given to citizen monitoring, participation, and transparency. The environmentalists had an effect, with EDF working with the EPA on the science side of the annex, and Kamp's group working on the public participation side. Thanks to La Paz and the opportunities it gave for networking among federal, state, and local officials and citizens' groups, the process of border diplomacy was changing from the old conventions, including the traditional Mexican preoccupation with sovereignty, into a cooperative mode. The agreement so casually signed by President Reagan at La Paz became the touchstone for a new regional partnership.

Not that accommodating the environmentalists was easy or comfortable. Howekamp recalls that Kamp's role was, basically, "to politicize the issue." For an air expert dealing with the nuts and bolts of policy, public diplomacy was sometimes hard to take. Metzger relished his role and his experience in government, but felt that "Kamp was a pain in the butt. He pushed hard on the Gray Triangle, he worried Babbitt to death, and he went to Washington and to Mexico City, and while effective, he didn't know when to stop."[35]

Perhaps the most important result of these interactions was to create personal contacts, familiarity with each other's problems, and new expectations, all of which began to build the new, close working relationship that exists today. "The La Paz working groups meeting every six months created a new relationship," Metzger reflected, which happened on the official side and also among the NGOs, starting in the late 1980s. Typical was the Grupo de los Cien, which was started in 1985 by artists, writers, and scientists in Mexico City to protest the government's neglect of land and water issues and soon moved to issues of national and transborder concern. As the 1990s dawned, a variety of Mexican NGOs were addressing environmental problems. But when Kamp went to Mexico in August 1984, he dealt with government officials and the press: there were no independent NGOs. Environmental groups did not escape the blandishments and dictates of government paternalism until the Mexico City earthquake.[36]

In fact, it must have been painful for environmentally minded Mexicans to realize how unprepared they were to negotiate. "SEDUE people were very cautious, sort of like small bureaucrats controlling information," Roberto Sánchez recalled, "and that hurt Mexico," which he felt should have gotten better terms. "SEDUE asked for very little information about Douglas from the United States—they had mostly what the Border Ecology Project sent them. They went along with the U.S. policy of saying, 'Douglas is our problem and we'll take care of [financing] it; you take care of Nacozari.'" That SRE's Alberto Székely did have the power to make financial decisions about financing pollution controls, whether an acid plant or sewage facilities, Sanchez acknowledged.[37]

What he did not know, of course, was that Central America was the immediate motivator. Nor did he perhaps appreciate that Székely had been attending meetings of the U.S.-Mexico Working Group, run by the cochairs, Ambassador Cesar Sepulveda and Professor Albert Utton of the University of New Mexico Law School. Set up in the mid-1970s with Rockefeller Foundation funding, this binational group of academics and government officials examined natural resource and environmental issues including water and air quality along the border. Székely himself strongly believed that past dealings with water and the environment were no longer adequate and that responses to isolated situations should be replaced with a more comprehensive view of problems. Needed, in short, was "a more holistic scope as the two countries consider the problems associated with their boundaries."[38] Given this perspective, La Paz and the smelter annex, both of which he was centrally involved in negotiating, were neither casual or reactive.

Having defined the Gray Triangle, Kamp was nothing if not assiduous in pressing on with *his* agenda. Funded by Robert Wick, the publisher, in the fall of 1983 the Smelter Crisis Education Project began to work as "a catalyst for an air pollution accord between the United States and Mexico. Specifically," he wrote, wearing both his hats as journalist and activist, "the accord would mean the development of an annex . . . that can control emissions from the three smelters to lesson their impact on such things as health, ecology, development, agriculture, acid rain, tourism and astronomical observatory development, while taking into account the economic impacts of pollution control." This multipronged initiative would include:

• On-site visits to smelters and meetings with mining officials on both sides;

• Frequent reports on socioeconomic developments to provide an equal information base to all;

• Organizing a binational conference, with EPA funding, to draft an accord;

• Direct consultation with government officials in both countries "to help formulate policy and increase intercommunication";

• Frequent talks with public and private groups in Arizona and elsewhere;

• Direct formulation and collection of data on emissions, meteorology, and health impacts; and

• Direct involvement in the corporate actions of Phelps Dodge to try to influence the company to reduce production at Douglas, once the Nacozari smelter is operating, until the current closure date for the Douglas smelter of January 1988."[39] The once-faltering Bisbee group had big ideas, and it is remarkable how much of this agenda was achieved.

To bolster the citizens' involvement part of this agenda, Wick arranged a meeting for Governor Babbitt, who heard a cross section of Sierra Vista citizens speak about the ill effects of Douglas pollution on visibility over the nearby Mule Mountains, which threatened the economic prospects of their rapidly growing city. This idea was Kamp's, and Babbitt took it all in. Wick recalls this session as the high point of his own engagement in the smelter issue.

Following another part of the agenda, Wick and Kamp met with senior Phelps Dodge officials at corporate headquarters in New York City, and then addressed the annual shareholders' meeting about the role of Douglas in the Gray Triangle. Wick, a former professor of art history and a sculptor, remembered the look of the New York office, with its intimidating deep green interior rooms that lay behind a reception area with a glass case displaying brilliant crystals and minerals. They briefed President George Monroe and two vice presidents, Richard Pendleton and Leonard Judd, about the Mexican situation, pointing out how, in Kamp's words, "the Mexicans view the pollution from Douglas as an impasse to an accord. We suggested to the PD officials that even planning to reduce production to one furnace from the current two furnaces would be a positive step that could free the diplomatic process." They also asked whether the corporation "was prepared to do anything to ease the environmental crisis, given that Douglas is under a 'death sentence' of January 1988, and, if funds were available, would Phelps Dodge consider providing technological assistance for environmental controls at Nacozari, since the uncontrolled Nacozari smelter is modeled after PD's sulfur-controlled Hidalgo, N.M., smelter?"

Monroe's response as reported by Kamp was candid. He and his associates considered Douglas to be integral to the success of their domestic operations. The company had been operating at a substantial loss because of depressed copper prices and competition from Third World countries backed by World Bank loans that were producing without pollution controls. Normally, Phelps Dodge ran about 25 percent of its concentrates through Douglas, which would now be handling more since the modifications at Morenci smelter were not panning out. Being cheaper to run, Douglas would operate for an extended period of time, processing some Ajo concentrate and Tyrone ores while Hidalgo was shut down for repairs. "Perhaps in 10 years, when the Tyrone, N.M., mine begins to diminish production, PD can think about voluntarily reducing production at Douglas, but at present, it is unthinkable, Mr. Monroe said." Furthermore, the company would continue to lobby for an NSO extension to operate until 1993. Finally, they could not offer any more expertise on pollution control to Nacozari than any contractor building an acid plant could provide.

Judd then asked Kamp, given this situation, which smelter he would rather see closed. "I answered uncomfortably, but truthfully, that my major initial motivation was to stop waking up to the Douglas plume choking me from 20 miles away while eradicating the Mule Mountains." Yet, Nacozari would be much larger and would impact the whole southwestern airshed, but eventually would be controlled. Kamp really did not know what it would be like to wake up in Bisbee Junction with the Nacozari plume coming from a tall stack sixty miles away. "I think," said Mr. Judd, "that you will be pleasantly surprised." And this "small exchange" was symbolic, Kamp concluded, "of the different worlds we live in. Separate realities. I did not really feel that patriotic U.S. smoke will really be more breathable than competitive Mexican smoke. 'We really don't feel that your concerns are all that justified,' Mr. Monroe said with a smile."[40]

Shareholders at the annual meeting the next day reacted with indifference to a debate between Monroe and Kamp on Douglas's environmental record. Owning one share of Phelps Dodge stock gave Kamp access, but his attempt to get a shareholder's resolution on the ballot for the next annual meeting was headed off by Phelps Dodge. Having tried and failed with a conciliatory approach, Kamp abandoned the dry hole of corporate democracy. For its part, the corporation made the mistake of not taking "that hippie from Bisbee" seriously.

Kamp's great success was in first establishing the Gray Triangle as a fact and then in operationalizing the concept through networking in the ways specified above. It bears repeating that he operated as a team with Robinson and Yuhnke. Pooling their various talents and orchestrating the system in a brilliant display of civic activism, this triumvirate was unstoppable.

They operated in a society where the climate of acceptance for mine and smelter practices had changed. Since pioneer days, mining had been seen as a positive good, producing value from nature. Now millions of environmentally conscious North Americans viewed mining as the rape of nature. Once perceived as corporate innovators, progressive in their town and labor management practices, PD had become an embattled corporation.

Priscilla Robinson, Bob Yuhnke, and Dick Kamp in front of the closed Douglas smelter, January 17, 1987. (Photo courtesy of Priscilla Robinson)

By their lights, of course, Phelps Dodge was in a war for corporate survival. Efforts to secure tariff relief from copper imports and an extension of the NSO crested in 1984 and did not clear the Senate. The effort to operate Douglas indefinitely ran up against two new realities: a stiffening regulatory climate and the fact that the EPA and the state of Arizona were now running on parallel tracks with respect to enforcement.

The EPA might be prepared to grant an NSO variance to the SIPs, but not without changes in the way the SCS at Douglas was being operated and a plan to capture fugitive (nonstack) emissions. On March 14, 1985, the corporation notified the EPA and Arizona of its intent to apply for a second-period NSO, and in May it submitted a proposal along with plans for a third generation of the SCS system and a design and work plan to deal with fugitive emissions.[41] After months of negotiations, the EPA rejected both proposals on technical grounds as inadequate to protect the NAAQS. Given this rejection, the requirement of a specific timetable for interim constant controls, such as an acid plant, was not addressed because PD now agreed to close Douglas at the end of this second-period NSO, on January 1, 1988.[42]

Pat Scanlon, the senior official in charge, recalls having difficulty pin-

Arizona governor Rose Mofford, in pink hard hat, at La Caridad for start-up of the new acid plant. (*Arizona Daily Star,* July 28, 1988, p. 17b; courtesy of the *Star*)

ning the EPA down on the specifics of this NSO. The company really tried to meet the terms of an NSO in 1985, working until midnight and on weekends to perfect their submission. However, "the NSO was interpreted in a way that was very rigid, whereas the law said that NSO's were predicated on a reasonable plan to protect the air quality." But then Phelps Dodge heard from the EPA people in San Francisco that the plan was not detailed enough to show how the company was going to do it. "EPA horsed us around to the point we never did get an NSO." Nor did PD get one the first time around for that matter, because the EPA was so slow promulgating the regulations. Basically, Scanlon said, the EPA saw the NSO as a sweetheart deal between Mo Udall and PD, "and they were determined to fix the bastards."

In the meantime, DRW operated on temporary permits granted by the state, which by 1985 was stiffening its own enforcement efforts. Phelps Dodge was not prepared to curtail operations further, which would have been necessary in order to comply with the particulate standard that the corporation had been challenging in court. Scanlon said that their tests of the instrumentation being used by the EPA to measure particulates in the stack gases of various smelters around the country showed it was not reliable. The company challenged this finding in court (see Chapter 5) and thought they had a good case.[43] But then PD was forced to accept the Arizona particulate standard (using the federal protocols) as a condition of the consent decrees signed in 1981 before Morenci and Ajo could receive operating permits. Absent an NSO for Douglas, they tried hard to get a state permit in 1986, but this request was denied on the ground of past violations. When Douglas was shut down, the consent decree negotiated with the EPA included a fine for past exceedences of the state particulate standard.

Officials at PD always felt that Governor Babbitt and the steelworkers union had a hand in the denial of this permit, acting in league with the environmentalists. Even before the strike, the United Steelworkers, based on research by Priscilla Robinson (who presumably got it from Kamp), had sponsored a bill in Congress requiring PD to put acid recovery technology in its stacks, "a requirement that would bankrupt the plant."[44] Poststrike, steelworkers linked up with the environmentalists—Robinson's grandfather had been a union organizer—and did what they could to make life more difficult for Phelps Dodge, claiming Douglas should be shut down for violations of health and environmental standards. In turn, PD argued that certain groups, such as the union and its allies, should be excluded from the state's permit hearings on the basis of harassment to the company. From the company's perspective, the campaign to close Douglas was payback time for the Morenci strike. Putting pressure on Phelps Dodge from the environmental side was just another way to hurt the company. In the emotions of the moment, this view seemed justified, but the pressures on the company were more broad based.

Not that Phelps Dodge lacked support—in the state legislature, for one thing, which in 1983 urged Congress to extend the life of the Douglas smelter for five years beyond 1988. Leaders of both Arizona houses also petitioned Secretary of State George Shultz and Lee Thomas not to support the position being taken by certain State Department officers that the EPA should deny the second NSO in order to "'preserve the United States Government's flexibility in the negotiations with Mexico.'"[45] The Reagan White House supported both these positions. Ever the pragmatist as well as the environmentalist, Babbitt appeared willing to support the NSO as long as Phelps Dodge committed unequivocally to shutdown in 1988 and took measures to protect Mexico.[46] In its quest for the NSO, PD responded in February with an olive branch for both Babbitt and the EPA, agreeing to enter into negotiations for a consent decree in federal court to ensure the definitive closure of Douglas by January 1, 1988, and in the interim to install SCS monitors south of the border to protect the air in Agua Prieta and surroundings.[47] In the Arizona congressional delegation, both

DeConcini and John McCain (who had Udall's old Tucson seat) supported the company's request for a waiver from the interim requirements to install continuous controls on Douglas.

But the regulatory process, while responsive to pressure, was on course. The fact that Douglas was now the last smelter operating in the United States without continuous sulfur dioxide controls prompted a tough response, pointing to shutdown before January 1, 1988. The company's failure to provide assurances on the technical side that the SO_2 NAAQS would be attained during the term of the NSO was one reason given for denial. The failure to submit a comprehensive plan for SO_2 abatement to the state was one reason for Arizona's denial of an operating permit.

The health issue was another factor, albeit one being held at arm's length, in the EPA's denial of a second NSO. They were in fact wary of EDF's petition, which they felt might have produced results inconsistent with EPA policy. Although Babbitt welcomed PD's conciliatory moves, he continued to insist that no NSO be granted until it could be shown that continued operation of the plant would not be detrimental to the health of asthmatics and other sensitive persons as well as comply with national air quality standards.[48] This stance was Yuhnke's doing as well as the fruit of discussions between officials at ADHS and Region 9 and the maturing regulatory process. EDF continued to press the health protection issue, and as has been described, the tape of June Hewitt and her asthma figured prominently in the EPA hearings held in Douglas that June. Soon after, the state declined to renew its permit to operate, forcing the plant to close early, in mid-January 1987.

For the EPA, the other compelling public policy argument was that issuing an NSO "would not provide assurance that the U.S. commitments to Mexico can be met." Bringing Douglas into compliance might require lengthy litigation beyond the target date for closing, by which date Mexico was promising to bring its two plants into control. Furthermore, Babbitt's argument in favor of reciprocity with Mexico was cited, along with his other concerns.[49] The July 1985 agreement with Mexico on the smelters was the catalyst that brought all of these strands together. Since 1983, many had worked to entwine the technical case against Douglas for noncompliance with the Mexican smelter issue. The Gray Triangle was a concept everyone could readily grasp. Now, with the prospect of a smelter annex under La Paz in sight, addressing the Gray Triangle menace as a whole was a public policy goal. What finally shut Douglas down was a framework that everyone could use and from which Phelps Dodge could find no extrication.

In a prepared statement, the company argued that the steelworkers union, the governor, and some elements of the environmental movement had been urging that Douglas be shut immediately in order to encourage Mexico to install pollution controls of their state-owned copper smelters. "This appeal, stripped of its rhetoric, constitutes a callous sacrifice of American jobs for no meaningful purpose." Mexico was using Douglas as an excuse to delay or avoid imposing controls on its two plants. Shutting down Douglas would not guarantee compliance. Given Mexico's financial straits, it was possible the new plant would not be finished on time, or the acid plant would be delayed indefinitely. Furthermore, "a second Mexican smelter at Cananea has operated for years with no controls whatsoever. Before demanding that Douglas be closed and Americans be placed out of work, it would seem logical to request that the Cananea smelter install an intermittent control system similar to Douglas, including production curtailments and a monitoring system designed to meet ambient air quality standards."[50]

Attempting to discredit or cast doubt on the Mexican linkage was a logical tactic to employ. There was no reason to welcome the onset of Mexican competition in an industry prone to overproduction and wide price swings. However, demonizing the United Steelworkers, a union that no longer represented PD workers and which it had thoroughly defeated in the bitter strike, was not good public relations. That Douglas was in compliance with the ambient air standards was an oft repeated line, but no longer credible. That the old plant would be closed rather than bear the cost of modernization was credible: almost four hundred well-paying jobs were on the line. On the defensive, Phelps Dodge failed to recognize that the changing times required a more creative response to the environmental challenge.

The EPA left the door open for Phelps Dodge to comply, and to get the NSO the company proposed a formula to the agency in February that included a consent decree, fines, and the placing of monitors in Mexico. But the EPA imposed stringent requirements to reach compliance, and it had the legal means to do so under the terms of the consent decree. Aside from further upgrading its SCS, which the EPA never approved, the only way to meet these requirements was to further curtail production, which would have made the plant unprofitable to operate.

In the letters that the EPA had been receiving from federal lawmakers about the Phelps Dodge NSO application, a common thread was the need to solve pollution in the Gray Triangle. This cause united lawmakers from such states as Colorado, where the worry was about acid rain, from Wyoming, where the state's lone congressman, Dick Cheney, was seeking advantages for low-sulfur western coal; and from the Arizona delegation, who were responding to constituent pressures to do something about pollution from the copper smelters.

As well, if local feeling ran high against Phelps Dodge, it must be said that there was no love lost for Mexico. Reciprocity was an idea that appealed to the public policy elite, including the NGOs, who were, nonetheless, quite willing and able to use the Mexican pollution threat when it suited their own purposes. But there was a "get Mexico" aspect to local feeling in Arizona that was worrisome and made some of the main players uneasy. To this extent, the way that the Gray Triangle issue was resolved was not altogether constructive; public opinion at the grass-roots level had not embraced cooperation.[51]

When it was all over Dick Kamp toured the new acid plant at Naco-zari, and in a letter to supporters he reflected on what this facility had cost Mexico financially, and how far that country still had to go:

I had the dubious honor of leading the U.S. environmental charge [Kamp said with forgivable exaggeration], feeling guiltier and more sullied every time some politico hurled another rock at Mexico for threatening the pristineness of the Rocky Mountains.

We leave the smelter with the usual sore glands and lungs, after leaving the manager of the smelter with a bottle of Chivas Regal and the acid plant manager some champagne to be opened when a stuck valve gets repaired. I leave hoping the same thing that I always hope for Mexico and its people—that things will keep working for them.[52]

All things considered, the outcome was more than bittersweet. Naco-zari is a state-of-the-art facility that continues to expand, most recently adding electroplating capability to recover 99 percent of slag resides, while continuing to build capacity in the workforce. Acquired by Grupo México, the successor to Mexicana de Cobre, Cananea was due to be shut down and merged with the efficient Nacozari operation. Under Annex IV, Cananea could continue operating without controls if it did not expand production. As for Phelps Dodge, it returned to profitability—thanks to the deep restructuring of the 1980s and the upswing of copper prices—becoming the largest copper producer in the United States. Douglas was dismantled; Cananea was living on borrowed time; and La Caridad de Nacozari, the one leg of the Gray Triangle with a modern plant, was still being monitored under the seminal La Paz agreement.

CHAPTER SEVEN

Continental Pathways

The North American mining and smelting industry has been thinking and acting continentally since the late nineteenth century, particularly so with respect to pollution issues, as the two smelters profiled in this book show well. Thus the industry developed networks and procedures to advance its common agenda decades ahead of the development of an adequate institutional response to the challenge of air pollution. For the mine and smelter interests, borders were no barrier to the fraternity of Canadian and American mining engineers and investors who developed the North American industry. Not so for governments at the federal and subfederal levels whose responses to the challenge of air pollution were subject to sovereignty constraints and the lack of a strong regulatory regime backed by an environmental consciousness. By the 1980s, borders were eroding for the environmental fraternity as well.

The Trail smelter and the associated mines were initially owned and operated by American interests, but by the early 1900s this industrial complex was controlled by the Canadian Pacific Railroad, its output protected by a tariff. Previously, lead ores had been smelted in Spokane, which was linked by rail to British Columbia. Then the industry was Canadianized, although Trail, like Canadian Pacific itself, was influenced by American management techniques and technology. Early on, the industry developed a tradition of sharing innovations, and senior engineers and managers moved readily between companies and across borders. This practice was abundantly clear in the case of Phelps Dodge, whose Arizona properties were developed by the Douglases from Montreal. Industry as a whole benefited from the advances in metallurgy and SO_2 recovery pioneered at Trail. In turn, the SCS dispersal system installed at Trail that became the control strategy of choice for industry was developed by two Americans, Professor Robert Swain of Stanford University and Reginald Dean of the U.S. Bureau of Mines, working with Canadian government meteorologists and Trail's own technical experts.

With respect to American trade and investment in Mexico, the border was meaningless until the 1911 revolution. Later, mine and smelter interests accommodated Mexican nationalism quite well, in contrast to the foreign oil companies whose assets were expropriated in 1938. The high McKinley tariff of 1890 and the imposition of new duties on the entry of lead ores into the United States simply made it profitable to build smelters across the Rio Grande. Cananea, an American company town, stimulated demand for American manufactured goods as well as jobs for American and Canadian engineers.[1] Sonora was pulled into the orbit of Arizona. American interests remained heavily invested until the 1970s when Mexicanization began. But here again the tradition of sharing technology and management techniques prevailed. The acid recovery plant at the new Nacozari smelter was financed and supervised by Canadians. The purchase of a used converter plant from Douglas enabled Cananea to expand production. Both Mexican complexes readily adopted the solvent extraction and electrowinning process (SX-EW) developed and perfected by Phelps Dodge engineers.

The conclusion must be that a shared industrial culture developed early on in the continental space from British Columbia to Sonora. The stratified towns were similar—from Rossland to Dawson to Cananea—or, at least, obvious similarities overlaid local and national differences. That this was not continentalism, as the term is understood in Canada, is clear. That it did foster a community of interests is also clear. Is it too much to identify this shared industrial culture as an early manifestation of the evolving North American community?

Until recently, no such claim could be made on the side of government. The development of a shared public policy to protect the environment is occurring only now, a century after the mine and smelter industry took on its continental shape. Indeed, as James Allum points out, policy in Canada and the United States supported and played into industry's successful attempt to reconstruct the narrative of an upper Columbia River landscape reclaimed from Indians and farmers who were identified as losers and on the way out. The future belonged to an industrialized landscape under the Trail smelter, which was the progressive and necessary successor. And while the farmers on both sides of the line received some small monetary relief, the upshot of the long Trail dispute was that industry retained the right to pollute private property as well as public lands across the border in the name of progress.[2]

To be sure, industry's victory was won at the cost of accepting a control regime. Together with the acid recovery plants and the fertilizer operation, the SCS developed for Trail was widely thought to have solved the smoke problem and the by-products problem. George Hedgcock, the veteran Forest Service investigator, said as much in his posthumous report published after the war with a collaborator. The landscape and the forests on both sides of the border had recovered, Hedgcock wrote, and in his opinion the effort to prove the invisible injury thesis was over.[3]

As mentioned, none of the American scientific reports by the USDA were published. The Canadian scientific work by the National Research

Council was published; *Effect of Sulphur Dioxide on Vegetation* was subsidized by Consolidated Mining and Smelting Company and distributed to the public. With respect to the hearings, the crisply argued Canadian critique of the American scientific case and the Canadian argument was published by Consolidated in a companion volume entitled *Trail Smelter Question.* This second volume was "given confidentially to those smelters, industries and other parties who were interested in the outcome of the case because certain principles of sulphur dioxide injury were being advanced which, if sustained, would have a material bearing upon sulphur dioxide nuisance litigation generally."[4] The contrast with the U.S. government's underfunded, divided, and ultimately bungled case is startling.

A generation later the balance of forces shifted toward a broader view of the public interest, thanks to the environmental movement of the 1960s, followed by the Clean Air Act of 1970 and the creation of the EPA one year later; new findings into the dangers of lead poisoning and the effects of SO_2; and the growing power of environmental NGOs using litigation to advance their views. Reinterpreting the landscape, many scientists saw smelting as a clear and present danger to nature as well as a threat to human health. As researchers learned more about what was in smelter smoke, the plumes that had streaked across western landscapes for generations were no longer tolerated as necessary adjuncts to production but as abominations creating pollution hellholes unsafe for people as well as agriculture and forests. Furthermore, in the copper industry after 1950, the flash furnace, a costly new Finnish technology that did not require venting smoke into the atmosphere, permitted the efficient recovery of smelting's by-products by using a closed-loop system. Enhanced recovery processes were developed for lead and zinc plants as well.

Armed with a new regulatory regime, the EPA succeeded in putting controls on the smelters, each one of which was easy to identify as a point-source polluter. (This policy was more difficult in the case of coal-fired power plants and automobile pollution, where the airshed as a whole was impacted, necessitating different strategies.) Controlling the smelters was a question of money, will, and the application of public policy interacting with corporations who accommodated to the new regime in various ways while striving to stay in business. All these issues are well illustrated in the case of Douglas. What lessons can be learned by comparing the two cases?

That the British Columbia–Washington border was linked to the Arizona-Sonora border in terms of technology and corporate culture in the mine and smelter industry is clear. By contrast, the institutional response was fragmented by borders and was slow to develop.

Consider that the Trail dispute began at the state-provincial level but was soon taken to the federal level for arbitration and resolution. So, too the Gray Triangle was resolved at the federal level, not only because Mexican law so required but also because the state of Arizona was subject to federal criteria mandated by the EPA and had been slow to act. However, recent developments at the state-provincial level draw into question the wisdom of always federalizing cross-border pollution disputes, which is not necessarily the best approach. In fact, finding the right level of government to handle disputes is a most interesting and pressing issue today. What does history tell us?

In the case of Trail, the province of British Columbia had jurisdiction over natural resources under Canadian law, but its role was superseded when the dispute became a foreign policy issue, taken first to the IJC and then to arbitration under a three-judge panel. The state of Washington was eager to turn it over to the State Department on the grounds that the purchase of property and smoke easements by the foreign corporation was barred by the state constitution. Blocked from receiving compensation by these means at home, the U.S. farmers felt all along that they would not receive fair treatment in Canadian courts (which they saw as pro-smelter in the treatment of Canadian farmers) and welcomed federal involvement. Nor was it likely that the British Columbian courts would have accepted jurisdiction in suits based on damage to land situated outside the province.

Aside from the scientific outcome, which can be viewed as stunted and constrained by the narrow legalism of the outcome, Trail is still of interest for affirming the "polluter pays" principle across international borders. In practice, this principle was quite favorable to industry, which was able to continue polluting over borders at small cost. Ironically, the precedent chosen by the Arbitral Tribunal came from the Ducktown smelter case, in which the U.S. Supreme Court affirmed that the Tennessee-based smelter owed compensation for damage to forests in the state of Georgia. U.S. laws were more favorable to industrial enterprise—"more evenly balanced in their effect on industrial and agricultural enterprise"—than the law of nuisance as set forth in precedents binding on Canadian courts, so American practice was selected.[5] In light of this evidence, there can be no question that the Trail smelter dispute as it has been traditionally taught in law schools needs to be reinterpreted by legal scholars.

In the late 1980s, citizens in the Northport area again complained about cross-border pollution from Trail—specifically the effects of air pollution and slag dumping into the Columbia River—and they fully expected it would be taken to the IJC for investigation. However, this time Washington State and British Columbia had an existing institution in place to deal with a range of cross-border issues, including oil spills, watershed contamination, and industrial pollution on the Columbia, in this instance slag dumping from Trail and the release of dioxin from paper mills up-river of the U.S. border. This organization was the British Columbia–Washington Environmental Cooperation Council, the origins and scope of which are analyzed by Jamie Alley in a recent essay.[6] The council accommodates provincial sensitivities in the Canadian West, where Ottawa is seen as being both remote and a meddler, and it also reflects the new-found willingness of American state governments to take on some responsibilities for cross-border environmental management.

The council is not a rule-making body, but it provides officials at the subfederal level with a way to share research and coordinate policies before disputes escalate into foreign policy problems. For example, when because of health concerns Trail was required to redesign its lead smelter, Washington State officials were asked to comment on the remediation plan. In turn, the state spent over $250,000 in a five-year investigation of long-standing local health concerns in the Northport area "of higher than normal incidences of everything from inflammatory bowel diseases to cancer, to thyroid disease, to respiratory diseases, to lead poisoning."[7] The state health department also investigated complaints about the effects of slag dumping into the Columbia, where the release of heavy metals was thought to be contaminating fish in Lake Roosevelt and endangering human health and animals by contaminants lodged in slag deposits along the banks. (According to the *Vancouver Sun*, "an estimated 15 million tonnes of slag line the riverbed and shoreline of the upper Columbia River.")[8] Although this investigation did not substantiate the worst of these concerns, Trail ended the century-long practice of slag dumping and pollution runoff from the plant.[9]

Recall that health and slag-dumping issues surfaced during the earlier investigation. In 1930, the U.S. Public Health Service was reluctant to conduct an in-depth study of health effects on the Northport community, alleging that such a small population could not produce statistically relevant results. As for the dumping, Stewart Griffin of the USDA wrote a paper on it that the Tribunal did not accept on the grounds that its charge was to investigate the effects of air pollution only. While it would be ahistorical to hold officials in the 1930s to the standards of today, when so much more is known about the effects on human health of SO_2, lead, cadmium, copper, and other heavy metals, it does seem in retrospect that these issues were finessed rather casually. Less charitably, these are additional examples of why the scientific investigation was stunted and something of a misfire.

As for the citizens, their hopes for a large award were bitterly disappointed when the 1937 hearing in Spokane revealed serious weaknesses in the U.S. case, but the smoke pollution was substantially reduced by the time the case closed four years later. In the 1990s, their concerns received prompt attention from the state and provincial authorities, who headed off involvement by the IJC that would have raised the profile of local complaints. In today's climate of antigovernment attitudes on the U.S. side of the line, the timely engagement of Washington state officials was not appreciated by everyone when the findings came out differently than many of the locals had expected. Studies of local health effects continue. And in 1996, agencies of the two governments formally agreed to cooperate on environmental issues in the Columbia River Valley. The council concluded: "When added to the steps the smelter has already instituted to eliminate discharge of slag, reduce contamination of site runoff, and change internal plant processes and waste streams, the prospects seem good that within the next few years COMINCO's discharge will meet BC Water Quality Criteria in the Columbia River at Trail."[10]

As for the council, it acts within a context of federal regulations and standards, weaker in Canada because of reduced enforcement by federal authorities due to cutbacks but still recognizably national. It can be concluded from this recent Trail example that subfederal actions based on cooperative approaches can work well in addressing cross-border environmental issues. But the council is not a rule-making body, and its legitimacy and credibility are imbedded in the regulatory regimes of two federal systems that uphold national standards. Currently, the council is suspended pending resolution of a fisheries dispute, but the agencies continue to cooperate and share information.

States along the U.S.-Mexico border are now becoming more active in environmental matters as officials at the subfederal level gain experience working together on local problems. Devolution of certain monitoring and enforcement responsibilities to the Mexican states will accelerate this process.

Larger, systemic problems in bilateral environmental management are clearly the province of federal authorities working in cooperation with subfederal governments and local groups. Regional issues such as cross-border air pollution and watershed contamination along the northern border are being addressed by the IJC. The IJC has proposed the establishment of permanent international watershed boards. In 1996, members of the IJC's International Air Quality Advisory Board went to Ciudad Juárez and El Paso to compare notes on airshed management on the other border. A trinational agreement on ways to monitor and measure airborne pollutant pathways across North America is not far away, although, to be sure, Mexico considers that management of the airshed overall is of more concern to the two industrialized nations.[11] Looking further ahead to global warming, will Mexico join with its two trade partners to support the 1997 Kyoto accord by accepting a cap on its emissions of greenhouse gases while participating in a joint implementation program of emissions trading permits to meet gas reduction targets? If so, the NAFTA nations would follow Europe as the second bloc to develop a regional strategy to abate global warming.

To raise such a question just a generation ago would have been a fantasy; today it is a serious possibility. Then, relations between Canada and Mexico were minimal, and although U.S. interests held the largest foreign investments in Mexico, federal officials in Mexico City and Washington did not know each other well. And while the two governments cooperated to staunch the hemorrhaging of Mexican national accounts in the 1982 debt crisis, Mexico and the United States had different regional policies toward Cuba and Central America. From this perspective, the 1983 La Paz agreement to address common border concerns is all the more innovative. It was La Paz that provided the framework to solve the environmental challenge of the Gray Triangle.

It bears repeating that smelter pollution is under control in the Rocky Mountain West from British Columbia to Sonora. On both borders it was done bilaterally. However, there was a spillover into continental perspectives as regulators and environmentalists shared experiences and points of view. One may ask how it could be otherwise, since the industry had long been organized continentally (as I define it). True, but only with the caveat that for the environment it was a long time coming. Now we are on the cusp of new trinational institutions and practices, some of them prefigured on the U.S. southern border as well. What are the lessons from Arizona-Sonora?

For one thing, Mexicanization of the mines and smelter installations at Nacozari and Cananea took place within a history of technology sharing in the industry. It is safe to say that the arrival of environmental concerns in the 1960s was susceptible to being seen as an industrywide challenge, not as yet another American imposition on Mexico, the victim of an unhappy past. For another, both presidents had their own reasons for signing the La Paz agreement, which was the result of adroit staff work by Alberto Székely and Cliff Metzger. La Paz was a great confidence builder. Officials learned to cooperate just as the environment was becoming a hot domestic issue for the first time in Mexico, now in fact in both countries.

In recognizing that no workable alternative to cooperation is available to solve cross-border environmental problems, La Paz is more important than the sum of its parts. Implementation under the various annexes has generally been slow to date. However, it may well be that La Paz will come to be seen as a key turning point in Mexican-U.S. relations that, in taking on the environment, now played out along a broader bandwidth than the already existing close economic and financial relationships.

Later, upholding La Paz and the spirit of cooperation was a main argument of the Mexicans in opposing the siting of a low-level nuclear waste dump at Sierra Blanca, Texas, within thirty kilometers of the border. In October 1998, the state of Texas denied a construction permit to this facility, which would have accommodated nuclear waste from Vermont and New Hampshire as well as from Texas. The favorable publicity garnered by the Mexican government for its stand on Sierra Blanca may help it to generate the necessary political support and resources to deal with the estimated 60 percent of all radioactive wastes and industrial toxics that are disposed of illegally in Mexico.[12] In both countries, the waste must go somewhere. In turn, Texas's unhappy experience attempting to deal with Sierra Blanca as a state regulatory problem may help ease the way in solving the federal-state jurisdictional issues that have caused Washington to hold up signing a transboundary environmental impact assessment (TEIA) agreement with Mexico and Canada.

Annex IV, the smelter agreement, has in fact worked well. Douglas was shut down, and Nacozari, a state-of-the-art facility and the third largest smelter in the United States and Mexico after the Kennecot smelter in Utah and the San Manuel smelter in Arizona, has complied with the monitoring and reporting functions mandated by the agreement. Nacozari has a permitted SO_2 emissions limit of .650 ppm averaged over a six-hour standard. Prior to the installation of a second contact acid plant in 1997, Nacozari's record of meeting this Annex IV stack limit on a monthly basis was inconsistent. Cananea has continued to operate without controls, which it was permitted to do when Annex IV was signed in 1987 as long as the smelter did not expand capacity. Today, it has the dubious honor of being the largest nonferrous smelting source of SO_2 in Canada, the United States, or Mexico, emitting approximately 390 metric tons per day of SO_2, but has been slated to shut down for several years, having been kept open in part to maintain jobs. In Mexican political iconography, Cananea is an important symbol of labor militancy before the great Revolution of 1911. The town is divided over whether the plant should be shut down for environmental health reasons or allowed to remain open to protect some 350 jobs at the smelter.[13] However, the tailings pond that polluted the San Pedro River into Arizona in the 1980s has been rebuilt: another spill would pollute into a different watershed, not a good solution but better than putting at risk a riparian corridor that supports the migration of birds between Canada, the United States, and Mexico. Meanwhile, on the American side, toxics seeping into groundwater from old tailings ponds in the Douglas area are a legacy of the DRW.

Annex IV also resulted from the persistent and effective pressure of NGOs, notably the Border Ecology Project and EDF, to codify reduced emissions targets for the Gray Triangle. This watchdog role was less effective ten years later when the specter of Mexican coal-fired power plants polluting into Big Bend National Park arose. Why did the two governments and the NGOs sit by as Carbón I and Carbón II were permitted to come on line without controls? The bilateral consultations under La Paz work well when the actions of interested officials and NGOs coincide. A case in point is the recently signed emergency response agreement between Brownsville and Matamoros, sparked by EDF under the rubric of Annex II. The existence of a flexible framework for cooperative environmental management like La Paz, now fifteen years old, does not by itself assure a timely and effective response to shared problems as they arise. It is available for new tasks when effective local advocacy is brought to bear. In 1996 federal officials signed off on the Paso del Norte air quality management agreement, the first local plan to monitor and clean up a shared airshed on either border.[14]

With respect to the siting of industrial facilities on both borders, the three nations are close to signing a transboundary environmental impact assessment (TEIA) agreement, if State Department lawyers agree. This agreement would assure that the environmental effects of a cross-border site could be studied, anticipated, and abated before production began. Hopefully, local people across the border will have standing in the permitting hearings. If TEIA is ratified, anything like the Trail smelter dispute or the Gray Triangle question would be addressed before environmental problems arose. The uncontrolled Carbón power plants would not be operating today.

It is true that many of the more pressing air pollution challenges are in boundary areas and call for binational cooperation and binational action. But what of the continental commons? "Although a number of national, binational, and even global instruments exist, the lack of a North American approach to the complex problem of air pollution is striking," a recent Commission for Environmental Cooperation report concluded. In some cases there is a need for trinational cooperation "to complement and strengthen binational initiatives."[15] The smelter cases show how specific bilateral issues intersected to become regionally important. From this perspective, it is high time to recognize a shared responsibility for airborne pollution in the region as a whole, because "acting alone, no one nation of North America will be able to protect adequately its domestic environment or its citizens from pollutants transported along continental pathways."[16]

Common air quality standards do not yet exist for North America, although each country has established its own set of guidelines and standards. In North America, the United States emits by far the largest amounts of airborne pollutants, including sulfur oxides (SO_2, SO_x), nitrogen oxides (NO_x), carbon monoxide (CO), particulate matter (PM_{10}), volatile organic compounds (VOCs), and lead (Pb). Canada accounts for much of the rest. Prevailing winds transport pollutants northward, so that pollutants released in Mexico often impact on the United States, affecting visibility and air quality, while pollutants released in the United States cause acid rain in Canada. For this reason, Mexico is relatively less interested in developing a continental air regime and more actively engaged in specific cross-border issues such as managing the shared airshed of the Ciudad Juárez–El Paso metroplex. Abating the use of DDT, which is an airborne pollutant coming north from Mexico and Central America, is another case in point.

With respect to emissions of sulfur dioxide and the acid rain precursors (sulfur and nitrous oxides), the trend is down in all three countries. Under the 1990 Clean Air Act, the United States committed itself to reducing annual SO_2 emissions to 9.05 million metric tons below the 1980 level, a goal that will be largely reached by the year 2000, thanks to the highly successful use of tradable permits in the electric utility industry, which accounts for roughly 72 percent of total SO_2 emissions. Consequently, acid deposition into Canada—a major irritant in the bilateral relationship during the 1980s—is being substantially reduced.

As in the United States, Canada's emissions reductions have occurred in part because of aggressive targets and strategies that include industrial process changes, installations of scrubbers, and fuel switching. However, emissions in Canada come from a variety of sources including, most revealingly, the smelting of ores (50 percent), power generation (20 percent), and other sources (30 percent). Smelter pollution from Ontario and Quebec circling down into the New England states continues to be a problem.[17]

It is time to affirm the obligations of partnership, including the sharing of best practices in pollution abatement, just as the smelter industry has been doing for years. NAFTA itself provides an umbrella for a wide range of groups and organizations to associate for continental purposes. La Paz opened up a new era of cooperation, based on trust and (at times) sustained interaction among officials at the federal and subfederal levels. The spillover effect of La Paz is important, leading not only to a framework for solving the Gray Triangle issue but also to new types of pragmatic problem-solving in other areas, such as air, border emergency services, and toxics. The British Columbia–Washington Environmental Cooperation Council is a useful model for bilateral decision-making at the subfederal level, and groups like the Western Governors Association are discussing a range of actions appropriate at the state-provincial level in our three nations. British Columbia and Washington State already have a working Environmental Impact Assessment in place—all the more reason the TEIA should be signed now.

On the scientific side, the smelter issue is now clearly historical, and as a result of good public policy and industry's response, the problem of single-source polluters in the nonferrous smelting industry is resolved.

The shift to a health and welfare–based standard widened the scope of public concern beyond the narrowly defined legal standard for damages that had favored the smelters. Science and the law were now yoked into a more equal and symbiotic relationship. Politically, what was new in the climate leading up to the Clean Air Act of 1970 was the federal government's commitment to sponsor research as part of a holistic approach to abatement rather than the old, reactive, plant-by-plant approach that was allowed to prevail at Trail.

Furthermore, after a great deal of activity in the 1960–1990 period, research into the effects of SO_2 on plants and forests, and acid research in general, is now mature. The public policy focus now has shifted to the reduction in total loadings rather than the prior concern with loadings at specific sources like smelters. Pollution from power plants has been reduced substantially by the carrot of incentives, like emissions trading, as well as by the stick of regulation. But the United States and Canada have not yet coordinated their air policies sufficiently to manage the continental airshed. Aside from addressing air pollution in specific areas, notably Mexico City, the policy challenge for Mexico is to become the first industrializing nation to sign a joint implementation agreement with the North, acting together with its partners the United States and Canada to address global warming.

Decreased public funding for scientific research in the 1930s had a devastating effect on scientists from the USDA. Declining budgets today may similarly impact efforts to understand what is happening in the airshed as a whole and to support common modeling and monitoring programs. Public-private partnerships are one solution to the funding challenge. Ironically, the smelters, the big winners in the Trail dispute, made very effective use of this device. Thanks to the deep pockets of Consolidated, a CPR subsidiary, Canadian researchers at the NRC outspent and outperformed their American counterparts at the USDA.

Last but not least, the role of public participation has been discussed at length in these pages. In the former case, U.S. farmers were outgunned by industry even though they had two federal departments on their side. (The Bureau of Mines favored industry.) Furthermore, the Northport farmers acted within the paradigm of a traditional smoke action. They had no EPA or Clean Air Act to bolster their case for relief and compensation before the law. Contrast this to the new types of grassroots activism so well demonstrated in the Douglas case. The greatly expanded space given to the public, including the right to know and to participate in monitoring and formulating policy while exercising the right to bring suit against the government for noncompliance with environmental laws, is also amply demonstrated in the case of Douglas.

The expanding role of citizen participation is important to the smelter story. Citizen access to institutions is a touchstone of environmentalism as it has developed in North America. In fact, the engagement of local communities in addressing continental problems like air pollution is fundamental. Partnerships with businesses, NGOs, and governments at all levels may well become the order of the day. But whether the flow of history continues in this direction is a hope, not a fact. If it becomes fact, it will do so as a result of learning to act like partners in a North American community.

Environmental history encompasses several views of the natural world, and this book is no exception. Nature does not appear as a protagonost. Rather, man's relationship to nature in the mine and smelter business and in the developing science and public policy of air pollution is my way to tell the story. Environmental history is a big tent where many can rub shoulders and debate. Personally, I am not attracted to the "wounded nature approach," where the only good human action is not to have acted at all. And anyone interested in mining history would have a difficult time being a deep ecologist. Rather, given my background and experience, it is the principle coming out of the 1992 Earth Summit in Rio de Janeiro that appeals and informs: to create wealth, with equity, without ruining the planet.

This book began with a main theme: how it was that the two realms of business and public environmental policy intersected and ultimately came together in North America. Although the span of action covers nearly a century, the continental significance of this story was not apparent until recently. Perhaps it is now obvious in our globalized world. However, history still matters, especially if one asks the right questions.

NOTES

PART I. THE TRAIL SMELTER DISPUTE

1. John D. Wirth, "The Trail Smelter Dispute: Canadians and Americans Confront Transboundary Pollution, 1927–41," *Environmental History Review*, Vol. 1, No. 2 (April 1996), pp. 34–51. For a succinct overview, see Mary-Lou Quinn, "Early Smelter Sites: A Neglected Chapter in the History and Geography of Acid Rain in the United States," *Atmospheric Environment*, Vol. 23, No. 6 (1989), pp. 1281–92.

2. James Robert Allum's dissertation, entitled "Smoke Across the Border: The Environmental Politics of the Trail Smelter Investigation" (Queens University, Kingston, ONT, 1995), is a well-written parallel study that reaches similar conclusions: Trail was a victory for industry. I obtained a copy of his work after writing my own narrative. Based largely on Canadian archives and the published hearings, Allum shows how industry "reimagined" the natural landscape to justify its dominance in a cross-border area that already had an aboriginal past and an agricultural present.

3. That smelters were liable for damages to crops and forests caused by air pollution was well established in Canadian and U.S. law. Litigants could settle individually or league together in a "smoke action" to hire their own lawyers and experts. But taking on corporate counsel and their in-house scientific experts was expensive, so they were under pressure to settle.

4. For a succinct account of the negotiations, see D. H. Dinwoodie, "The Politics of International Pollution Control: The Trail Smelter Case," *International Journal*, Vol. 27, No. 2 (Spring 1972), pp. 218–35. The negotiations, as well as the published hearings, are well covered by Allum, "Smoke Across the Border."

5. The phytotoxic effects from single point-source pollution on nearby vegetation are now well documented, although the complex interaction of several factors on plant morphology is still not yet fully understood.

6. This line of argument was suggested by my reading of Alfred P. Rubin's excellent discussion of the legal aspects in "Pollution by Analogy: The Trail Smelter Arbitration," *Oregon Law Review*, Vol. 50 (1971), pp. 259–82. See also Karin Mickelson, "Rereading *Trail Smelter*," *Canadian Yearbook of International Law*, Vol. 31, Tome 31 (1993), pp. 219–33.

CHAPTER 1. SETTLERS AND SMELTERS IN THE COLUMBIA RIVER VALLEY

1. Northport Over Forty Club, *Northport Pioneers: Echoes of the Past from the Upper Columbia Country* (Colville, WA: Statesman-Examiner, 1981), pp. 2, 4, 22, 102. James Robert Allum, "Smoke Across the Border: The Environmental Politics of the Trail Smelter Investigation" (Ph.D. diss., Queens University, Kingston, ONT, 1995), pp. 76–81.

2. Trail Smelter Arbitral Tribunal, "Decision . . . of April 6, 1938," typewritten

in CNA, RG 25, pp. 8, 10. Also A. E. Richards of USDA, "Northeastern Washington Economic Survey," in Trail Smelter Arbitral Tribunal, *Trail Smelter Question*, Documents Series, AA Economic Survey, appendix AA1 (Ottawa, ONT: Government Printing Office, 1937).

3. Richards, pp. 175, 176. In his sympathy for the farmers and their ordeal under smelter smoke pollution, Allum exaggerates their economic prospects (if not their courage in taking on the smelter).

4. R. V. Stewart, letter to state relief supervisor, Colville, March 3, 1934, in *Northport News* (March 16, 1934).

5. A. G. Langley, Report to the [BC] Minister of Mines, December 3, 1927, in USNA, RG 54, Records of the Office of the Chief Chemist, entry 149a, box 3.

6. Letter, Stewart Griffin to Dr. W. W. Skinner, Northport, November 4, 1928, in USNA, RG 97, Skinner file, entry 91, box 48. *Northport News* (January 22, 1935). Ed Morris to USDA, Northport, March 27, 1940, in USNA, RG 54, Records of the Office of the Chief Chemist, entry 149a, box 4.

7. William Dietrich, *Northwest Passage: The Great Columbia River* (New York: Simon and Schuster, 1995), p. 179, and Blaine Harden, *A River Lost: The Life and Death of the Columbia* (New York: Norton, 1996), pp. 106–7.

8. Elsie G. Turnbull, *Topping's Trail: The First Years of a Now Famous Smelter City* (Vancouver, BC: Mitchell Press, 1964), p. 37ff. "The years of 1905–06 saw the triumph of Walter H. Aldridge in his effort to establish a smelting industry in Canada and to ensure that the centre of that industry should be in Trail rather than in other smelter towns" (p. 58). Also Lawrence S. Whittaker, "All is not Gold," unpublished company history dated November 19, 1945, in Add. Mss. 2500, COMINCO, BC Provincial Archives, pp. 102–3, 188.

9. Professor Robert E. Swain, letter to John Read, Palo Alto, CA, September 15, 1943, in CNA, RG 77, National Research Council "Trail Smelter Investigation," volume 93.

10. Blaylock started with Consolidated in 1900 as a chemist and joined Aldrich's management team. He served as assistant general manager under J. J. Warren, becoming managing director at Warren's death in 1938, then president and finally chairman before his own death on November 19, 1945. See Whittaker's excellent profile in "All Is Not Gold."

11. Lon Johnson, "Memoirs," typewritten, Colville, 1963, in Stevens County Historical Society, p. 35. Stevens County Superior Court, civil court cases 6542 and 6600, microfilm 146, 147. The Canadian suit is detailed in *An Appeal from the Court of Appeal for British Columbia Between Consolidated Mining and Smelting Company Defendant and Alfred Endersby Plaintiff*, Victoria, 1919, in add. mss. 2500, box 432, BC Provincial Archives. Also, Turnbull, *Topping's Trail*, p. 57.

12. John Leaden et al., petition "To the Honorable Governor and Legislature of the State of Washington," from Northport, December 1, 1925, files of Governor Hartley, Washington State Archives, Olympia.

13. D. F. Fisher, "Report on Injury Caused by Smelter Smoke in Northern Stevens County, Washington," typewritten, Wenatchee, WA, August 12, 1927, in USNA, RG 54, entry 13b, box 1.

14. *The Deans' Report: Final Report to the I.J.C. by Dean E. A. Howes and Dean F. G. Miller*, pp. 12–13, September 1929; published as appendix A1 of the *Trail Smelter Question* (Ottawa, ONT: Government Printing Office, 1936).

15. Johnson, "Memoirs," pp. 52–53. Leaden to Blaylock, Northport, April 6, 1929, recounting efforts to discredit the CPA, in Washington State Archives, Hartley correspondence. Affidavit by the three Stevens County commissioners, dated June

2, 1927, in CNA, RG 25, vol. 1495, file 578-A-27 I, and their resolution of June 9, 1928, in the Hartley correspondence. Professor Edward C. Johnson, dean and director of the State College of Washington, made a long report at the secretary of agriculture's request: Letter, Johnson to Jardine, August 9, 1927, in USNA, RG 54, Bureau of Plant Industry, Bureau Chief's correspondence, entry 2, file 1600, box 490.

16. Letter, Daniel Landon, chair of the Appropriations Committee, to John Leaden, December 18, 1925, in Washington State Archives, Hartley correspondence. Letter, Griffin to Skinner, November 4, 1928, in USNA, RG 97, Skinner file, entry 91, box 48, which included a clipping from the *Daily Province*, a Canadian paper that reported that "a smelter as large as that at Trail could be operating at full capacity continuously for years beyond the lifetime of the average man now living." Also, Letter, Griffin to Skinner, October 18, 1928, Skinner file.

17. R. C. Crowe, statement of November 23, 1927, to A. G. Langley, engineer at the BC Ministry of Mines, included with supporting documents in a memorandum of February 27, 1928, by O. D. Skelton to Washington, asking whether—as the State Department requested—an IJC reference was really necessary if the Washington State land laws could be amended; in USNA, RG 54, entry 149a, box 3. Also, Leaden, letter to Blaylock, April 6, 1929, in Hartley papers.

18. *Northport Pioneers*, Leaden family profile on pp. 210–15. Interview with Dorothy Leaden, March 1995. Theodore Saloutos and John D. Hicks, *Agricultural Discontent in the Middle West, 1900–1939* (Madison: University of Wisconsin Press, 1951), p. 112, and Ira E. Shea, *The Grange Was My Life* (Fairfield, WA, 1983).

19. Letter, Leaden et al. to Blaylock, Northport, April 6, 1929, Washington State Archives, Hartley correspondence.

20. Incidents reported by the USDA's Goldsworthy to D. F. Fisher, from Northport on May 10, 1930; W. S. Boyer of the Interior Department, writing to the commissioner of the General Land Office in Washington, July 27, 1928: Report of a meeting of the Citizens' Protective Association, when a crowd of some four hundred people met with the governor, state representatives, and a state senator at the Isis Theatre on June 22; both in USNA, RG 54, Bureau of Plant Industry, Bureau Chief's correspondence, entry 2, file 1600, smelter smoke, box 490. The Blaylock death threat letter, dated June 9, 1934, is in CNA RG 25, vol. 1690, file 103-II.

21. Warren to John E. Read, Montreal, June 14, 1934, with copy to the prime minister, CNA, RG 25, vol. 1690, file 578-A-27 I. Warren to Mackenzie King, written from his vacation home in North Carolina, December 3, 1927, and Warren to O. D. Skelton, undersecretary of state for External Affairs; both in RG 25, vol. 1495.

22. Raftis was born in nearby Chewelah, WA, of Canadian-American parents, and his wife Loretta was from Ontario; Leaden's daughter-in-law Dorothy was from Rossland across the line. Sketch in *The People Who Will Live in Colville Area History* (Colville, WA: Statesman-Examiner, 1979), p. 120. John T. Raftis, *The Raftis Family of County Kilkenny* (Colville, WA, 1968).

23. Memorandum for the prime minister, Trail Smelter, Ottawa, July 9, 1928, in Skelton papers, CNA, RG 25. Dr. Goldsworthy to K. F. Kellerman, Wenatchee, Washington, in USNA, RG 54, Bureau of Plant Industry, Bureau Chief's correspondence, entry 2, file 1600, box 490.

24. R. V. Stewart, letter to state relief supervisor. Turnbull, *Topping's Trail*, pp. 67–68.

25. Governor Hartley to Attorney General John Dunbar, Olympia, January 28, 1926, in Washington State Archives, Hartley correspondence. The quote is from a memorandum by Knight to Paul Appleby, March 27, 1938, in USNA RG

97, Skinner file, entry 91, box 48. Letter, Minister William Phillips of U.S. Embassy to W. L. Mackenzie King, secretary of state and prime minister, Ottawa, December 22, 1927, in CNA, RG 25, vol. 1495, file 578-A-27 I.

26. Jamie Alley, "The BC/Washington Environmental Cooperation Council: An Evolving Model of Canada/US Interjurisdictional Cooperation," in Richard Kiy and John D. Wirth, eds., *Environmental Management on North America's Borders* (College Station: Texas A&M University Press, 1998).

27. Skelton to J. H. King, minister of health, Ottawa, December 27, 1927, CNA, RG 25, D1, vol. 741, file 148, on microfilm. Dispatch, U.S. Minister Phillips to Secretary, Ottawa, July 12, 1928, in USNA, Department of State, RG 59, decimal file 711.4215, air pollution/75, box 6597.

28. Memorandum, Wilbur J. Carr to Jacob Metzger and Mr. Barnes (State Department attorneys) on his joint testimony to the Bureau of the Budget with Dr. K. F. Kellerman, associate chief of the Bureau of Chemistry and Soils, February 15, 1928, in USNA, Department of State, RG 59, decimal file 711.4215, air pollution/48, box 6597. Also, Henry Knight Diary, vol. 1, November 7, 1929, pp. 162–63, in USDA Library, Beltsville, MD.

29. Memorandum, Griffin to Skinner, February 23, 1939, in USNA, RG 54, Records of the Office of the Chief Chemist, entry 149a, box 3.

30. Knight Diary, entries in vol. 1 for July 11 and July 17, 1929, pp. 93 and 97.

31. Letter, Skinner to Griffin, Washington, November 6, 1928, in USNA, RG 97, Skinner file, entry 91, box 49.

32. Skinner to Griffin, Washington, November 3, 1928, in USNA, RG 97, Skinner file, entry 91, box 49.

33. Telegram, Warren to Magrath, head of the Canadian section of the IJC, Trail, April 30, 1929, in Skelton papers, CNA, RG 25.

34. Letter, Griffin to Skinner, Northport, September 16, 1929, and letter, Griffin to Skinner, May 14, 1930, both in USNA, RG 97, Skinner file.

35. Memorandum, "Conference Between Doctors Knight, Kellerman, McCall, Skinner, and Mr. [James Oliver] Murdock relative to future investigation," March 6, 1930, in USNA, RG 97, Skinner file, box 51.

36. Note, Vincent Massey, Canadian minister in Washington, to Secretary of State Stimson, May 1, 1929, and reply to State Department by R. W. Dunlap, acting secretary of the Bureau of Plant Industry, on May 16, 1919, in USNA, Records of the Office of the Chief Chemist, entry 149a, box 1. Skelton to C. A. McGrath, head of the Canadian section of the IJC, Ottawa, June 26, 1929, in CNA, RG 25, vol. 1495, file 578-A-27-I.

37. Letter, Griffin to Skinner, Northport, November 5, 1929, in USNA, RG 97, Skinner file.

38. Letter, K. F. Kellerman to Hedgcock, Washington, May 29, 1930, in USNA, RG 54, Bureau of Plant Industry, Bureau Chief's Correspondence, entry 2, file 1600, smelter smoke, box 490.

39. Memorandum, Griffin to Skinner, October 24, 1938, in USNA, RG 97, Skinner file, entry 91, box 48; letter, Griffin to Skinner, May 14, 1930, entry 91, box 49; the Chairman (Judge Hostie), 1940 hearings, in Trail Smelter Arbitration, *Hearings*, vol. 54, pp. 6188–89.

40. Sidney Norman, "Smelter Fumes Involve U.S. Sovereignty," *Globe and Mail* (August 5, 1937).

41. Letter, D. F. Fisher to Dr. K. F. Kellerman, chief, Bureau of Plant industry, from Wenatchee, WA, November 14, 1929, in USNA, RG 54, Bureau of Plant Industry, Records of the Chief, entry 2, file 1600, smelter smoke, box 491.

42. [Morris Katz et al.] *Effect of Sulphur Dioxide on Vegetation* (Ottawa: National Research Council, 1939). Also Morris Katz, "Sulfur Dioxide in the Atmosphere and Its Relation to Plant Life," *Industrial and Engineering Chemistry*, Vol. 41, No. 11 (November 1949), pp. 2450–64.

43. An excellent summary of this research is by Theodore C. Scheffer and George G. Hedgcock, "Injury to Northwestern Forest Trees by Sulfur Dioxide from Smelters," USDA, Forest Service, *Technical Bulletin No. 1117* (June 1955). The map is on p. 22; emission figures are on p. 14.

44. George G. Hedgcock, "Smelter Fumes Injury to Crops in Northern Stevens Country, Washington, in 1936," p. 23, typewritten, in USNA, RG 54, Bureau of Plant Industry, Correspondence of the Chief, entry 2, file 1600, smelter fumes, box 491.

45. Scheffer and Hedgcock, "Injury to Northwestern Forest Trees," p. 43.

46. Griffin to Skinner, April 18, 1938, "Salient Points, Concepts and Considerations of the Trail Smelter Award by the International Joint Commission," in USNA, RG 97, Skinner file, entry 91, box 48.

47. S. G. Blaylock, vice president and general manager, to the Arbitral Tribunal, eight pages, typewritten [1936], in CNA, RG 25, vol. 1690, file 103-III. J. N. Robinson, "The History of Sulfur Dioxide Emission Control at COMINCO Ltd.," *International Journal of Sulfur Chemistry*, Part B, Vol. 7, No. 1 (1972), 52–53.

48. Robinson, "The History of Sulfur Dioxide Emission," 51.

49. Robert E. Swain, "Atmospheric Pollution by Industrial Wastes," *Industrial and Engineering Chemistry*, Vol. 15, No. 3 (March 1923), p. 301. Letter, Oldright to Diamond, Salt Lake City, September 19, 1929, in USNA, RG 70, USBM General Classified Files 482.1, box 791.

50. Memorandum for the prime minister, July 26, 1928, in Skelton papers, CNA, RG 25.

51. Knight Diary, February 5, 1930, vol. 1, p. 209. Letter, Goldsworthy to Senator W. L. Jones, from Wenatchee, April 7, 1931, and to his boss, K. F. Kellerman, on March 28, 1931, in USNA, RG 54, Bureau of Plant Industry, Bureau Chief's correspondence, entry 2, file 1600, box 490. Memorandum for the prime minister, February 28, 1930, in Skelton papers, CNA, RG 25.

52. Letter, Griffin to Skinner, Northport, March 9, 1931, in USNA, RG 97, Skinner file, box 49.

53. Letter, Goldsworthy to Kellerman, March 28, 1931, in USNA, RG 54, file 1600, box 490.

54. Skelton to Herridge, the minister in Washington, from Ottawa, October 31, 1934, Skelton papers, CNA, RG 25. Letter, Warren to Read, Montreal, April 7, 1934, in CNA, RG 25, vol. 1690, file 103–34.

55. Letter, Crowe to Read, Trail, May 3, 1934, in CNA, RG 25, vol. 1690, file 103-34.

56. Memorandum by the legal adviser, March 18, 1935, detailing Metzger's conversation with the CPA directors on March 8 in Spokane, in USNA, RG 76, Records of Boundary and Claims Concerns, Trail Smelter Arbitration, Reports of Scientific Investigators, entry 31, box 6. Telegram, William Phillips (acting secretary of state) to Governor Clarence D. Martin, requesting that "every facility be extended to the Canadian investigators" [A. W. McCallum, the forest expert; the NRC's Dr. Ledingham, the plant pathologist; and Morris Katz, the chemist], June 4, 1934, in Washington State Archives, Clarence Martin papers.

57. Alfred P. Rubin, "Pollution by Analogy: The Trail Smelter Arbitration," *Oregon Law Review*, Vol. 50 (1971), pp. 261, 264, 270.

58. Read's informative account, "The Trail Smelter Dispute," appeared in *Canadian Yearbook of International Law*, Vol. 1 (1963), pp. 213–29.

59. Metzger died on February 28, 1937. He was respected by Knight, who commented in the diary that Metzger had carried the U.S. case in his head, which left Swagar Sherley, the new agent, in the dark. Threatened by "certain men in the smelting industry," Metzger had "decided to go on with the case anyway job or no job" (Knight Diary, May 1, 1937, vol. 1, p. 1002–3). From Northport, Griffin said how much "he held the confidence and esteem of the people here," in USNA, RG 97, Skinner file, entry 91, box 49.

60. IJC, Trail smelter reference, statement of G. L. Oldright of the USBM, typewritten, Washington, D.C., April 2, 1929, in COMINCO, add. mss. 2500, box 432, BC Archives, especially p. 35. Oldright's supportive testimony (on April 22, 1929) was cited by Blaylock in his 1936 statement on the progress of his program.

61. Letter, Dean to Bureau Chief Finch, June 9, 1937, in USNA, RG 70, USBM general classified files 482.1, box 791.

62. Dean to A. P. Feldner, from Ottawa where hearings were taking place, October 14, 1937, in USNA, RG 70, USBM general classified files 482.1, box 791. Bureau policy was to treat information provided by the companies as privileged unless authorized by them to release it.

63. Letter, Raftis to Metzger, from Colville, February 6, 1937, in USNA, Department of State, RG 59, decimal file 711.4215, air pollution/743^{1}/$_{2}$. Leaden lingered until April 25, 1938.

64. Memorandum for Mr. Paul H. Apelby, March 28, 1938, in USNA, RG 97, Skinner file, entry 91, box 48.

65. Typewritten "Abstract of Publication," for release with the 1944 bulletin, in USNA, RG 97, Skinner file.

CHAPTER 2. "THE MOST ILLUSIVE OF PROBLEMS":
THE SCIENTIFIC DISPUTE

1. Report of the Selby Smelter Commission in J. A. Holmes et al., *U.S. Bureau of Mines Bulletin* 98 (Washington: Government Printing Office, 1915). "This most illusive of problems" is George Hill's comment in a letter to W. W. Skinner from Salt Lake, June 19, 1929, in USNA, RG 97, Skinner file, box 51.

2. Letter, D. F. Fisher, a USDA plant pathologist who conducted some of the first studies of forest damage in the Northport area, to Dr. Brooks at the Bureau of Plant Industry, Wenatchee, WA, July 15, 1929, in USNA, RG 97, Skinner file, box 51.

3. Letter, G. L. Oldright to R. S. Dean, the USBM's leading expert in smoke remediation, Salt Lake, June 24, 1936, in USNA, RG 70, USBM general classified files 482.1, box 791.

4. Letter, R. C. Crowe to Read, Trail, BC, May 3, 1934, in CNA, RG 25, vol. 1690, file 103.

5. Arthur Johnson, Swain's graduate student, summed it up in his 1932 dissertation: "no very conclusive results in regard to 'invisible injury' have yet been obtained by this procedure" ("A Study of the Action of Sulphur Dioxide at Low Concentrations on the Wheat Plant, with Particular Reference to the Question of 'Invisible Injury'" [Ph.D. diss., Stanford University, 1932], pp. 16–17).

6. Robert E. Swain, "Atmospheric Pollution by Industrial Wastes," *Industrial and Engineering Chemistry*, Vol. 15, No. 3 (March 1923), p. 298.

7. O'Gara tables in Moyer D. Thomas, "Effects of Air Pollution on Plants," World Health Organization monograph series no. 46, *Air Pollution* (Geneva: WHO, 1961), pp. 236–37.

8. Having talked with Hill and Thomas, Griffin reported: "It seems, however, that there is much of O'Gara's work that Mr. Thomas, the chemist now actively in charge under Hill, does not consider as reliable or worthy of publishing." Letter to Skinner, from Northport, July 18, 1929, in USNA, RG 97, Skinner file, box 49. Letter summarizing O'Gara's results, Hill to R. C. Crowe, Salt Lake, July 13, 1929, in USNA, RG 97, Skinner file, box 51.

9. Letter, Griffin to Skinner, Northport, July 18, 1929, in USNA, RG 97, Skinner file, box 49.

10. Letter, Hill to Skinner, Salt Lake, June 19, 1929, and Long report, Fisher to Brooks from Wenatchee, July 15, 1929 in USNA, RG 97, Skinner file, box 51. George R. Hill and M. D. Thomas, "Influence of Leaf Destruction by Sulphur Dioxide and by Clipping on Yield of Alfalfa," *Plant Physiology*, Vol. 8 (1933), pp. 223–45.

11. Profile by Eric Hutchinson, *The Department of Chemistry, Stanford University, 1891–1976* (Stanford, CA: Stanford University Press, 1977), pp. 8, 11–13. Robert E. Swain and Arthur B. Johnson, "Effect of Sulfur Dioxide on Wheat Development," *Industrial and Engineering Chemistry*, Vol. 28, No. 1 (January 1926), p. 47.

12. D. F. Fisher, "A Summary of Experimental Work and Observations on the Effect of S0$_2$ on Plant Foliage (1927–1930)," typewritten [1932?], p. 4, in USNA, RG 54, Records Relating to the Trail Smelter Investigation, 1927–1934, entry 13b, box 1.

13. Sorauer's third edition was translated by Frances Dorrence and published by the Record Press (Wilkes-Barre, PA: 1922). The definitions are quoted on p. 721 of this edition. In listing this German bibliography (in "Early Smelter Sites: A Neglected Chapter in the History and Geography of Acid Rain in the United States," *Atmospheric Environment*, Vol. 23, No. 6 [1989], p. 1284), Mary-Lou Quinn writes that "rarely is this [earlier] German effort ever mentioned in retrospective literature on acid rain."

14. H.-G. Dässler and S. Börtitz, *Air Pollution and Its Influence on Vegetation* (Dordrecht, Netherlands: Junk Publishers, 1988), citing Wislicenus's 1898 and 1914 publications, pp. 11, 45, 52.

15. Ibid., pp. 47–48, 51–52.

16. Julius Stoklasa, *Die Beschädingungen der Vegetation durch Rauchgasses und Fabrikexhalationen* (Berlin: Urban and Schwarzenberg, 1923), introduction, p. xxiv, and pp. 81–83, 104. Also, Dässler and Börtitz, *Air Pollution*, p. 51.

17. Swain and Johnson, "Effect of Sulfur Dioxide on Wheat Development," pp. 42–43.

18. Report of Commissioner Robert E. Swain to the U.S. District Court, Salt Lake City, Utah [1920], pp. 86–87, 101, in USNA, RG 70, USBM general classified files 482.1, box 790.

19. Ibid., p. 101.

20. John Raftis testimony in Trail Smelter *Hearings*, vol. 54, p. 6178.

21. Ibid., pp. 6185–86. He quoted at length from Swain's 1923 lecture, "Atmospheric Pollution by Industrial Wastes," pp. 296–301.

22. D. F. Skinner, "Reconnaissance of Smelter Injury, Stevens County, Washington, July 10–12, 1928," typewritten, in Skelton papers, CNA, RG 25, vol. 1495, file 578-A-17 I.

23. Fisher, "A Summary of Experimental Work," p. 12, in USNA, RG 54, entry 13B, box 1.

24. Skinner to Griffin, December 11, 1928, in USNA, RG 97, Skinner file, box 49.

25. The visit to Trail is in Fisher's report, p. 24. Fisher (1888–1949), who both read and spoke German, had an MA in horticulture. His record as an expert on fruit diseases, marketing, and transportation was outstanding. Skinner to Griffin, May 22, 1930, in USNA, RG 97, Skinner file, box 49.

26. Letter, Fisher to Kellerman, from Wenatchee, November 14, 1929, in USNA, RG 97, Skinner file, box 51.

27. Letter, Griffin to Skinner, July 18, 1929, in USNA, RG 97, Skinner file, box 49.

28. Fisher, "A Summary of Experimental Work," p. 3, in USNA, RG 54, entry 13b, box 1.

29. Letter, Griffin to Skinner, Northport, July 17, 1929, in USNA, RG 97, Skinner file, box 49: "The procedure employed at Salt Lake for testing the atmosphere for sulfur dioxide and the fumigation experiments in progress there interested me greatly." Also, Griffin to Skinner, July 18, 1929, in USNA, RG 97, Skinner file, box 49.

30. Griffin, "Summary of . . . Chemical Aspects of the Investigation, 1928–1930," typewritten [January, 1932?], in USNA, RG 54, Reports and Other Records of S. G. Griffin, entry 149c, box 15.

31. Ibid.

32. Fisher, "A Summary of Experimental Work," in USNA, RG 54, entry 13B, box 1.

33. L. R. Leinback, "Determination of Total Sulfur Content and Ash Content of foliage in the Smelter Area," typewritten, April 11, 1930, in USNA, RG 97, Skinner file, box 52.

34. Fisher, "A Summary of Experimental Work," pp. 26–27, in USNA, RG 54, entry 13B, box 1.

35. Leinback, "Determination of Total Sulfur Content," p. 35, in USNA, RG 97, Skinner file, box 52..

36. Griffin, "Summary of Smelter Fumes Investigation," in USNA, RG 54, entry 149c, box 15; emphasis added.

37. "Preliminary statement by the Smelter Fumes Committee analyzing the scientific factors embodied in the Report of the International Joint Commission . . . ," typewritten, March 29, 1931, in RG 54, entry 2, file 1600, box 490.

38. NRC, "Proceedings of the First, Second, Third, and Fourth Meetings of the Associate Committee on Trail Smelter Smoke," p. 1, Ottawa, 1929–1931, CNA.

39. The Canadians delayed moving their machine to Marble until Griffin chose his site at nearby Evans. "Our Marcus machine records sufficient gas to make it probable that traces could be found for great distances down the river. To record these traces will only help to extend this problem further into debatable ground and anything that can be done to keep Griffin in the upper valley would be beneficial"; George H. Duff, "Memorandum re Wenatchee Fumigation Experiment," p. 10, in NRC, "Third Meeting of the Associate Committee on Smelter Fumes Investigation, August 27, 1930," CNA. Difficulties in securing an underwater cable hookup from the C.M.&S. controlled power plant in Northport delayed the placing of a third U.S. recorder near the border.

40. Griffin to Skinner, Northport, May 14, 1930, and Skinner's reply of May 22, in USNA, RG 97, Skinner file, box 49.

41. Arbitral Tribunal, decision of April 16, 1938, pp. 20–21, in CNA, RG 25, vol. 1692, file 103A.

42. R. C. Crowe, ed., *Trail Smelter Question* (Trail, BC: Consolidated Mining and Smelting Company, 1939), preface.

43. Covering letter to the memorandum, Duff to F. H. Lathe, Northport, August 19, 1930, "Third Meeting of the Associate Committee on Trail Smelter Smoke," CNA.

44. Henry C. Knight, signing for the Smelter Fumes Committee's "Preliminary Statement" on the IJC's February 28 report, typewritten, March 17, 1931, in RG 54, entry 2, file 1600, box 490.

45. Duff memorandum, August 19, 1930, CNA.

46. Letter, Goldsworthy to Kellerman from Wenatchee, September 29, 1939; letter, Murdock to Goldsworthy from Washington, October 16, 1930; and memorandum from Fisher to Kellerman, October 17, 1930, all in RG 54, Bureau Chief's Correspondence, entry 2, file 1600, smelter fumes, box 490.

47. Letter, Skinner to Griffin, December 10, 1930, in USNA, RG 97, Skinner file, box 49.

48. Proceedings of the Fourth Meeting of the Associate Committee on Trail Smelter Smoke, February 14, 1931, CNA. Also Morris Katz, "Memorandum on the present and future status of the Trail smelter smoke investigation," August 11, 1931, CNA.

49. Katz, memorandum of August 11, 1931.

50. Memo, K. F. Kellerman, associate bureau chief, to Knight, July 25, 1928, quoting Dr. Metcalf in their budget bureau, in USNA, RG 54, entry 17, smelter fumes, box 2637. The actual expenditures are from a memo from McCall to Jump, March 31, 1938; in RG 54, entry 2, file 1600, smelter fumes, box 491, which contains Knight's observation as well.

51. Fourth meeting of the Associate Committee on Trail Smelter Smoke, CNA; Elsie G. Turnbull, *Trail Between Two Wars: The Story of a Smelter City* (Vancouver, BC: Morriss Printing, 1980), pp. 69–71.

52. Proceedings of the Eighth Meeting of the Associate Committee on Trail Smelter Smoke, October 8, 1935, including the quote from a letter by S. P. Eagleson, secretary-treasurer, to Read, September 10, 1935. Also, McNaughton to J. J. Warren, October 9, 1935, summarizing the discussion: "At a meeting held yesterday, fear was expressed that this procedure [of direct billing via expense vouchers] might make it appear to our American friends, if they became aware of it, that this investigation was being dictated by the company and was not in fact an impartial investigation of the NRC."

53. Read replying to Eagleson, September 23, 1935, and McNaughton's letter to Warren, source in footnote 46.

54. Knight wrote on p. 274 in his diary for May 22, 1930: "Mr. Metzger . . . informed us that the present act appropriating $40,000 for this research by the Department of Agriculture must be done on projects approved by the State Department" (Henry Knight Diary, USDA Library, Beltsville, MD).

55. Knight testimony before the House Committee on Appropriations, *State Department Appropriation Bill*, 72d Cong., 1st sess., February 1932, p. 274.

56. Knight Diary, December 15, 1932, pp. 594–92. Hill continued to be a staunch ally until he left the House in 1937. The loss of both Hill and Metzger was keenly felt by Knight, who then had to deal with the incompetent Swagar Sherley; diary for May 1, 1937, pp. 1002–3.

57. Knight Diary, March 18, 1932, p. 511.

58. Assistant Secretary Wilbur Carr, memorandum of July 15, 1932, and Metzger's reply of July 18, in USNA, Department of State, RG 59, decimal file 711.4215, air pollution/388.

59. Carr testimony, p. 267; Knight testimony, p. 288; Congressman Blanton, in House Appropriations Committee, 72d Cong., 0st sess., Report 1890, January 21, 1933, p. 289.

60. Letter, Skinner to Griffin, June 20, 1934, including his statement before the House Appropriations Committee on June 12, 1934, in USNA, RG 97, Skinner file, box 49.

61. Metzger, memorandum for Carr, February 26, 1935, in USNA, Department of State, RG 59, decimal file 711.4215, air pollution/586, and testimony before the Senate Appropriations Subcommittee, State, Justice, and Labor Departments Appropriation Bill, 74th Cong., 1st sess., February 19, 1935, p. 2.

62. Letter, Subcommittee Chairman Senator Kenneth McKellar to Cordell Hull, February 21, 1935, in USNA, Department of State, RG 59, decimal file 711.4215, air pollution/586, and Metzger's testimony, State, Justice, and Labor Departments Appropriation Bill, February 19, p. 3.

63. Senate Appropriations Subcommittee, April 1936, pp. 9–14.

64. Memorandum, Knight to Paul H. Appleby, May 6, 1938, in USNA, RG 16, General Correspondence of the Office of the Secretary, entry 181, smelter fumes (1938), box 2872; long memorandum, Knight to Appleby, March 28, 1938, in RG 97, Skinner file, box 48. Knight Diary, May 5, 1938, p. 1077.

65. Proceedings of the Eleventh Meeting of the Associate Committee on Trail Smelter Smoke, November 23, 1937, CNA.

66. Knight Diary, May 22, 1931, pp. 412–13; "Report of the Meeting of the Smelter Fumes Committee," same date, in USNA, RG 97, Skinner file, box 49; and letter, Griffin to Skinner from Wenatchee, June 25, 1931, RG 54, Records of the Office of the Chief Chemist, entry 149a, box 1.

67. The article appeared in volume 24 (August 1932). Letter, Skinner to H. E. Howe, the editor, March 3, 1932, and abstract, both in USNA, RG 97, Skinner file, box 52; Moyer D. Thomas to Howe, August 13, 1932, and Griffin's reply to Howe, September 3, 1932, Skinner file, box 49.

68. Knight Diary, January 6, 1934, p. 714. Memorandum, Henry G. Knight to Skinner, McCall, and Kellerman, June 10, 1934, in USNA, RG 54, Records of the Chief Chemist, entry 149a, box 1. Letter, Metzger to Knight, March 24, 1934; memorandum, Fisher to E. C. Auchter, January 22, 1934, in RG 54, Bureau Chief's Correspondence, file 1600, smelter fumes, box 490.

69. Memorandum, Griffin to Skinner, April 24, 1939, in USNA, RG 54, Bureau of Chemistry and Soils, entry 182, file 36240, box 148; letter, Hedgcock to G. A. Ledingham of the NRC, July 9, 1942, in CNA, RG 77, vol. 91, file 17-13T-3; Theodore C. Scheffer and George G. Hedgcock, "Injury to Northwestern Forest Trees by Sulfur Dioxide from Smelters," USDA, Forest Service, *Technical Bulletin No. 1117* (June 1955).

70. Proceedings of the Tenth and Eleventh Meetings of the Associate Committee on Trail Smelter Smoke, October 22 and November 23, respectively, CNA.

71. Statement, June 12, 1934, in USNA, RG 97, Skinner file, entry 91, box 49.

72. "Some months showed conditions to have improved in comparison to 1930—other months showed aggravated conditions. In March 1933, a fumigation lasting 57.67 hours was reported with a maximum concentration of 0.82 p.p.m. Reports subsequent to March 1933, indicate that there have been numerous fumigations of long duration with high concentration." Katz, "Trail Smelter Case,"

[early 1934], memo for Whitby and Lathe, in CNA, "Morris Katz Chemistry Special Reports, 1932–1944," box 5.

73. Letter to Swagar Sherley, August 11, 1937, in USNA, RG 97, Skinner file, entry 91, box 53.

74. George Hedgcock, "Report of Scientific Investigations in 1936," dated February 15, 1937, p. 18, in RG 54, Correspondence of the Chief, file 1600, smelter fumes, box 491.

75. R. C. Crowe, "Cumulative, Progressive Injury," in *Trail Smelter Question* (Trail, BC: Consolidated Mining and Smelting Company, 1939), pp. 99–101.

76. Knight Diary, April 6, 1937, p. 996, and May 4, 1937, pp. 1004–5. Having once been considered for vice president, Sherley was too prominent politically for the Smelter Fumes Committee to protest his appointment as U.S. arbiter.

77. Letter, Read to Tilley, December 28, 1936, in CNA, RG 25, vol. 1690, file 103-III.

78. R. S. Dean and Robert D. Swain, "Report of the Technical Consultants to the Trail Smelter Arbitral Tribunal: Meteorological Observations Near Trail, B.C. 1938–1940," typewritten, p. 374, in BC Provincial Archives, add. mss. 2500, box 482. For the atmospheric factors and methods of control, see pp. 345ff.

79. Memorandum, Griffin to Skinner, October 24, 1938, in USNA, RG 97, Skinner file, box 48.

80. Dean and Swain, "Report of the Technical Consultants," typewritten. Griffin's reports for 1939 and 1940 "showed that vegetation in the area was still being subjected to definitely harmful fumigations"; in Benedict English's covering memo, September 25, 1940, included in the "Supplementary Statement." Both documents are in the BC Provincial Archives, add. mss. 2500, COMINCO, box 482.

81. Letter, Hull to Ickes, June 12, 1941; memo, R. R. Sayers, director of USBM, to Michael Straus at the State Department, May 17, 1943, in USNA, RG 70, USBM general classified files 482.1, box 791. The section on plant damage contained several passages after page 273 affirming that the area was free of any indication of injury by SO_2. The final, revised report appeared without these passages in 1944 as *USBM Bulletin 453*.

82. F. H. Chapman, memorandum to A. L. Johannson, head of the NRC legal office, July 3, 1944, and letter, Read to Lathe, August 17, 1944, in CNA, RG 77, NRC, "Trail Smelter Investigation," vol. 91.

83. Letter, Swain to Read, Palo Alto, CA, September 15, 1943, in CNA, RG 77, NRC, "Trail Smelter Investigation," vol. 91.

84. The results are summarized in National Research Council, *Twenty-first Annual Report, 1937–1938* (Ottawa, ONT: NRC, 1938), pp. 118–22, and in the last chapter of the book, *The Effect of Sulfur Dioxide on Vegetation*.

85. Morris Katz, "Sulfur Dioxide in the Atmosphere and its Relation to Plant Life," *Industrial and Engineering Chemistry*, Vol. 41, No. 11 (November 1949), p. 2464.

CHAPTER 3. PATHS TAKEN AND NOT TAKEN

1. "Fume Troubles in the Offing," *Chemical and Metallurgical Engineering*, Vol. 44, No. 4 (April 1937), p. 179.

2. Letter, R. C. Crowe to Read, Trail, BC, May 3, 1934, in CNA, RG 25, volume 1690, file 103.

3. See Alfred P. Rubin's excellent discussion of the legal aspects in "Pollution by Analogy: The Trail Smelter Arbitration," *Oregon Law Review*, Vol. 50 (1971), pp. 259–82. Also see Karen Mickelson, "Rereading *Trail Smelter*," *Canadian Yearbook of International Law*, Vol. 31, Tome 31 (1993), pp. 319–33.

4. The suggestion was rejected by the tribunal and the two scientific advisers on the grounds that "such a regime would unduly and unnecessarily hamper the operations of the Trail Smelter and would not constitute a 'solution fair to all parties concerned'" (Trail Smelter Arbitral Tribunal, *Decision of the Tribunal, Reported March 11, 1941* [Washington, DC: Government Printing Office, 1941], p. 47).

5. Morris Katz memo or conversation with R. C. Crowe in Toronto, February 24, 1934, in CNA, RG 24, vol. 1690, file 103-I. Read, p. 6483, and Tilley, p. 6486, in Trail Smelter *Hearings*, vol. 56.

6. Letter to Sherley, August 11, 1937, in USNA, RG 97, Skinner file, entry 91, box 49.

7. Letter, Griffin to Skinner, Northport, May 24, 1937, in USNA, RG 97, Skinner file, entry 81, box 49.

8. R. E. Swain and R. S. Dean, technical consultants, memorandum on operating requirements to begin April 15, 1940, in CNA, RG 97, vol. 90, file 17-13T-3.

9. C.M.&S. Company, "Smoke Control Regime," pamphlet for employees excerpting the operating regime from the tribunal's decision of March 11, 1941.

10. Letter, F. H. Chapman to Ledingham, Trail, October 14, 1941, in CNA, National Research Council, "Trail Smelter Investigation" file, RG 97, vol. 91.

11. Long memo, "Re: Smelter Smoke Investigation," Lathe to Mackenzie, August 11, 1942, in CNA, National Research Council, "Trail Smelter Investigation" file, RG 97, vol. 91.

12. Letter, Lathe to Chapman, January 15, 1943, and July 14, 1943, in CNA, National Research Council, "Trail Smelter Investigation" file, RG 97, vol. 91.

13. George G. Hedgcock, "Smelter Fumes Inquiry to Crops in Northern Stevens County, Washington, in 1936," in USNA, RG 54, Bureau of Plant Industry, Correspondence of the Chief, entry 2, file 1600, box 491; Theodore C. Schaeffer and George G. Hedgcock, "Injury to Northwestern Forest Trees by Sulfur Dioxide from Smelters," USDA, Forest Service *Technical Bulletin* 1117 (June 1955).

14. Walter W. Heck, "Assessment of Crop Losses from Air Pollutants in the United States," in *Air Pollution's Toll on Forests and Crops*, ed. James J. MacKenzie and Mohamed T. El-Ashry (New Haven, CT: Yale University Press, 1989), p. 255.

15. Moyer D. Thomas, "Effects of Air Pollution on Plants," in World Health Organization monograph series no. 46, *Air Pollution* (Geneva; WHO, 1961), pp. 234–35, 240. Thomas C. Hutchinson, "The Ecological Consequences of Acid Discharges from Industrial Smelters," in *Acid Precipitation: Effects on Ecological Systems*, ed. Frank M. D'Intri (Ann Arbor, MI: Science Publishers, 1982), p. 116. H.-G. Dässler and S. Börtitz, *Air Pollution and Its Influence on Vegetation* (Dordrecht, Netherlands: Junk Publishers, 1988), pp. 47–48. Sagar V. Krupa, *Air Pollution, People, and Plants* (St. Paul, MN: American Phytopathological Society, 1997), p. 103.

16. E-mail, Schramm to John Wirth, May 9, 1996. Thomas Kluge and Engelbert Schramm, "Vom Himmel hoch, Eine Geschichte der TA Luft," *Kursbuch*, Vol. 96 (1989), p. 94.

17. Gerd Spelsberg, *Rauch Plage: Zur Geschichte der Luftverschmutzung* (Aachen: Köner Volksblatt Verlag, 1988), especially pp. 146–47, 150, 161ff. With narrower standards of evidence, German farmers and foresters lost more and more smoke damage cases and became discouraged (p. 151).

18. Ibid., p. 171.

19. C. F. Halliday, "A Historical Review of Atmospheric Pollution," in World Health Organization monograph series no. 46, *Air Pollution* (Geneva: WHO, 1961) p. 29.

20. Heck, "Assessment of Crop Losses," p. 255.

21. Kurt Noack's 1925 publication on SO_2's effect on chlorophyll production was the most recent German work cited in the bibliography of Arthur Beers Johnson's doctoral dissertation, "A Study of the Action of Sulphur Dioxide at Low Concentrations on the Wheat Plant, With particular Reference to the Question of 'Invisible Injury'" (Stanford University, 1932), p. 64.

22. Ray E. Neidig, "Summarized Report on the Condition of Trees, Shrubs, and Crops in the Columbia River Valley for the year 1944," in CNA, National Research Council, "Trail Smelter Investigation," RG 97, vol. 91.

23. Letter, F. E. Lathe of NRC to Dean, Ottawa, June 2, 1944, and letter, Dean, now assistant director of USBM, to the Honorable R. H. Carlin of Sudbury, Ontario, from Washington, December 6, 1945, in CNA, National Research Council, "Trail Smelter Investigation," file, RG 97, vol. 91.

24. Communication from Tony Hodge of the Canadian National Roundtable to John Wirth.

25. For a succinct account of the role of linkage politics in these negotiations, see D. H. Dinwoodie, "The Politics of International Pollution Control: The Trail Smelter Case," *International Journal*, Vol. 27, No. 2 (Spring 1972), especially pp. 229–32.

26. Metzger's comments in a meeting with Knight and Kellerman on December 10, 1932, where they discussed the danger of losing the appropriation for smelter fumes research. Henry Knight Diary, p. 593, in USDA Library, Beltsville, MD.

27. Letter, Morris Katz to Dr. G. S. Whitby, director, Division of Chemistry, NRC, Ottawa, April 16, 1934, in CNA, RG 25, vol. 1690, file 103-I. Morris Katz, "Some Aspects of the Physical and Chemical Nature of Air Pollution," in World Health Organization monograph series no. 46, *Air Pollution* (Geneva: WHO, 1961), p. 113.

28. S. G. Griffin, USDA senior chemist, writing to Lawrence Burchard from Northport, WA, June 5, 1937, suggesting how the counterargument could be made, in USNA, RG 54, Records of the Office of the Chief Chemist, NC 135, entry 149a, box 2. Also note 124, O. D. Skelton, undersecretary of External Affairs, to W. D. Herridge, Ottawa, June (?) 1934, in CNA, RG 25, vol. 1690, file 103-I. John E. Read, "The Trail Smelter Dispute," *Canadian Yearbook of International Law*, Vol. 1 (1963), pp. 224–28.

29. Bennett to Robbins, February 17 and 22, 1934, cited by Dinwoodie, "The Politics of International Pollution Control," note 38, p. 230.

30. Crowe to Read, May 3, 1934, and Skelton to the Montreal Coke and Manufacturing Company, Ottawa, May 9, 1934, both in CNA, RG 25, vol. 1690, file 103-I.

31. Ibid.

32. Stephen Blank et al., "North American Business Integration," *Business Quarterly*, Vol. 58, No. 3 (1994).

33. Aldridge and Guess are both profiled in *The National Cyclopaedia of American Biography* (Clifton, NJ: James T. White, 1984), vol. 43, pp. 90–91, and vol. 35, p. 400, respectively. The quote is from Donald Chaput and Kett Kennedy, *The Man from ASARCO: A Life and Times of Julius Kruttschnitt* (Parkville, Australia: Australian Mineral Heritage Trust and Australasian Institute of Mining and Metal-

lurgy, 1992), p. 81. Beckett's bio is in *Who's Who in America* (Chicago: A. W. Marquis, 1946), vol. 24, p. 161.

34. See "Canada and the American Institute," *Engineering and Mining Journal*, Vol. 115, No. 18 (May 5, 1923), p. 787, and Wilkins, *The Emergence of Multinational Enterprise: American Business Abroad from the Colonial Era to 1914* (Cambridge, MA: Harvard University Press, 1970), pp. 118–19.

35. *Phelps Dodge — A Copper Centennial, 1881–1981*, special supplement in *Pay Dirt* (Summer 1981), p. 52.

36. Stanford's new chemistry faculty first received national recognition for the pioneer work of Swain, Pierce, Mitchell, and Harken on smelter pollution, notably at Anaconda (1907) and Selby (1913). "Annual Report of the President of Stanford University for the Academic Year Ending August 31, 1925," *Stanford University Bulletin*, 5th series, no. 3 (January 1, 1926), p. 163. The fellowships continued while Swain chaired the department, until 1940. Letter, President Wilbur to Swain, Palo Alto, July 25, 1925, in Stanford University Archives, Wilbur correspondence, 1925–1926, SC39, box 2, folder 10.

37. Paul M. Ambrose, assistant chemist, Metallurgical Division, "Sulphur Dioxide in Smelter Smoke: A Review Based on Recent Personal Observations," restricted report prepared in 1936 in Tucson, AZ, in USNA, RG 70, USBM general classified file 482.1, box 790.

38. Ibid., p. 47. The photo appears as figure 6.

39. The record of industry pressure tactics is in USNA, Department of State, RG 59; for example, Metzger's memo of September 23, 1936, decimal file 711.4215, air pollution/705, box 4013. The ASARCO letter threatening his job is mentioned in his long memo dated February 12, 1936 (document 663).

40. Letter, Anderson to the Secretary, June 11, 1936, and Secretary's reply, in USNA, Department of State, RG 59, decimal file 711.4214, air pollution/686, box 4013.

41. Memo from the legal adviser (Metzger) to Hackworth, September 23, 1936, in USNA, Department of State, RG 59, decimal file 711.4215, air pollution/705, box 4013.

42. Letter, W. B. Gentry, assistant to the president of Freeport Sulphur Company, to General A. G. L. McNaughton, president of the NRC and others there, New York, January 9, 1940, in CNA, NRC, "Trail Smelter Investigation," RG 97, vol. 90.

43. Metzger memo on conversation with lawyers for the Texas Gulf Sulfur Company, February 12, 1936, in CNA, NRC, "Trail Smelter Investigation," RG 97, vol. 90.

44. Letter, Senators Morris Sheppard and Tom Connally, to the secretary, February 17, 1936, USNA, Department of State, RG 59, in decimal file 711.4215, air pollution/664.

45. Long memo to Hackworth, February 21, 1936, 10-11, in 711.4215 Air Pollution/664.

46. Letter, Chandler P. Anderson to Secretary of State Cordell Hull, Washington, D.C., March 24, 1936, in 711.4215 Air Pollution/674.

47. Read, "The Trail Smelter Dispute," p. 228.

48. Michael Hart, *Decision at Midnight: Inside the Canada-U.S. Free-Trade Negotiations* (Vancouver: University of British Columbia Press, 1994), and Gordon Ritchie, *Wrestling with the Elephant: The Inside Story of the Canada-U.S. Trade Wars* (Toronto: Macfarland, Walter, and Ross, 1997).

49. Read's devastating and detailed assessment of the hearings is in a letter

to undersecretary Skelton, marked "secret" from Spokane, July 22, 1937, in CNA, RG 25, O. D. Skelton Papers, vol. 741, file 148, microfilm 13.

50. Read, "The Trail Smelter Dispute," pp. 225–26.

51. Letter, Crowe (from Trail) to Tilley (in Toronto) April 27, 1938, in CNA, RG 25, vol. 1691, file 103-IX.

52. Letter, Crowe to Read, from Trail, April 22, 1938, in CNA, RG 25, vol. 1691, file 103-IX. Crowe sent copies of the award to Salt Lake and to the general counsel of Aldrich's Texas Gulf Sulphur Company.

53. Letter, Crowe (from Trail) to Tilley (in Toronto), April 27, 1938, in CNA, RG 25, vol. 1691, file 103-IX.

54. Ford Burkhart, "Gustave Freeman, 88, Scientist Who Studied Effects of Pollution," *New York Times* (September 28, 1987), p. 18.

55. Ridgway H. Hall, "The Lost Legacy of the Trail Smelter Case" (Honors thesis, Yale University, 1989), p. 9.

56. FPA meeting reported in *Northport News*, May 28, 1928. Opening brief by lawyers from Northport and Colville on behalf of the farmers, whose claims are set forth in appendix 2, in BC Archives, COMINCO, add. mss. 2500, box 475.

57. Report to the Surgeon General, October 17, 1930, in USNA, Department of State, RG 59, decimal file 711.4215, air pollution/163, and Smelter Fumes Committee discussions for November 6 and December 10, 1930, Knight Diary, pp. 342a, 358.

58. Miller and Howes, "Trail Smelter Investigation, Final Report," typewritten [1930], in RG 76, entry 31, Records of Boundary and Claims Commissions, Reports on Scientific Investigations, box 2.

59. Halliday, "A Historical Review of Atmospheric Pollution," pp. 28–29.

60. Letter, Griffin to Skinner, June 28, 1929, and June 8, 1930, in USNA, RG 97, Skinner, entry 91, box 49.

61. Christian Warren, "The Silenced Epidemic: A Social History of Lead Poisoning in the United States Since 1900" (Ph.D. diss., Brandeis University, Waltham, MA, 1997), pp. 380–401. Also Patterson's obituary in the *New York Times* (December 8, 1995), p. B18.

62. "Tests Relating to Pollution of the Waters of the Columbia River by Effluents from the Trail Smelter," one-page summary in a report on the experimental work conducted in FY 1932, sent to Skinner from Wenatchee, September 9, 1933, in USNA, RG 54, Records of the Office of the Chief Chemist, entry 149a, box 1.

63. Arbitral Tribunal, "*Decision . . .* of April 16, 1938," typewritten, p. 40, in CNA, RG 25, vol. 1692, file 103A. R. C. Crowe, ed., *Trail Smelter Question* (Trail, BC: Consolidated Mining and Smelting Company, 1939), part 1, p. 161, which briefly summarizes Griffin's unpublished report.

64. Estimate reported by Terry Glavin, "Toxic Chemical Worries Straddle the Border," *Vancouver Sun* (November 19, 1991), p. B13. Crowe reported that some 14,000 tons of slag each month were being dumped in the late 1930s (*Trail Smelter Question*, part 1, pp. 161–62).

65. "Chemical Character of the Waters of the Columbia River in Northeastern Washington," brief summary in memorandum for Skinner, December 15, 1937, in USNA, RG 97, Skinner file, entry 91, box 48. Blaine Harden, *A River Lost: The Life and Death of the Columbia* (New York: Norton, 1996), pp. 117ff.

66. International Joint Commission, Trail smelter reference, *Brief Submitted on Behalf of His Majesty's Government in Canada* (Ottawa, ONT: Government Printing Office, 1930), p. 7. While calling for a just solution, the report added: "It would not be to the advantage of the two countries concerned that the industrial effort should be prevented by exaggerating the interest of the agricultural community.

The Canadian government has a vital interest in establishing a precedent that will preserve an even balance between the conflicting interests" (in BC Provincial Archives, COMINCO, add. mss. 2500, box 445).

PART II. THE GRAY TRIANGLE CONFRONTATIONS

1. Lynton Keith Caldwell, "Binational Responsibilities for a Shared Environment," in *Canada and the U.S.: Enduring Friendship, Persistent Stress,* ed. Charles F. Duran and John H. Sigler (New York: Prentice-Hall, 1985), p. 218.

2. Sources for the corporate history include Robert Glass Cleland, *A History of Phelps Dodge, 1834–1950* (New York: Alfred A. Knopf, 1952), and *Phelps Dodge— A Copper Centennial, 1881–1981,* special supplement to *Pay Dirt* (Summer 1981). For the industry see George H. Hildebrand and Garth L. Mangum, *Capital and Labor in American Copper, 1845–1990: Linkages Between Product and Labor Markets* (Cambridge, MA: Harvard University Press, 1992).

3. Paul M. Ambrose, "Sulphur Dioxide in Smelter Smoke: A Review Based on Recent Personal Observations," Tucson, AZ [1936], in USNA, RG 70, USBM general classified files 482.1, box 790.

CHAPTER 4. SMOKE FROM DOUGLAS

1. Of a total employment of 1,871,875 jobs, only 10,531 remained in metal mining. Thomas Michael Power, *Lost Landscapes and Failed Economies: The Search for a Value of Place* (Washington, DC: Island Press, 1996), p. 98.

2. George H. Hildebrand and Garth L. Mangum, *Capital and Labor in American Copper, 1845–1990: Linkages Between Product and Labor Markets* (Cambridge, MA: Harvard University Press, 1992), pp. 159–61. Alecia Swasy, "Long Road Back; How Phelps Dodge Struggled to Survive and Prospered Again," *Wall Street Journal* (November 24, 1989), pp. A1, A3.

3. Power covers this topic well (*Lost Landscapes and Failed Economies,* pp. 102–12).

4. Jeremy Mouat locates Rossland in the "western cordillera" of mines and mining towns stretching, presumably, down into Mexico. The larger continental connections are implicit in his excellent study of a Canadian place and time in *Roaring Days: Rossland's Mines and the History of British Columbia* (Vancouver: University of British Columbia Press, 1995); the quotes are on pages 150 and 165. For a brief discussion of the two Mexican smelter towns and the 1906 Cananea strike, see Héctor Aguilar Camín, *La frontera nómada: Sonora y la Revolución Mexicana* (Mexico, DF: Cal y arena, 1997), pp. 150–51, 152–65.

5. Toby Smith, *Coal Town: The Life and Times of Dawson, New Mexico* (Santa Fe, NM: Ancient City Press, 1993).

6. Diana Hadley, "Border Boom Town—Douglas, Arizona, 1900–1920," *Cochise Quarterly,* Vol. 17, No. 3 (Fall 1987), pp. 3–47. For Douglas and its Mexican hinterland, see Josiah McC. Heyman, *Life and Labor on the Border: Working People of Northeastern Sonora, Mexico, 1886-1986* (Tucson: University of Arizona Press, 1991).

7. John Blackburn, "Future Bright for Unity, Growth of Douglas and Agua Prieta" (July 21, 1987), and "Douglas Weathers Most Effects of Smelter Shutdown" (December 31, 1987), both in *Sierra Vista Herald–Bisbee Daily Review.*

8. See especially Hildebrand and Mangum, *Capital and Labor in American Copper, 1845–1900*. Histories include James W. Byrkit, *Forging the Copper Collar: Arizona's Labor-Management War of 1901–1921* (Tucson: University of Arizona Press, 1982), and Jonathan D. Rosenblum, *Copper Crucible: How the Arizona Miners' Strike of 1983 Recast Labor-Management Relations in America* (Ithaca, NY: ILR Press, 1995). See also Mouat, *Roaring Days*, pp. 71–108.

9. Robert Glass Cleland, *A History of Phelps Dodge, 1834–1950* (New York: Alfred A. Knopf, 1952), pp. 165–69.

10. "One Hundred Years of Smelting at Trail," *Canadian Mining Journal*, Vol. 117, No. 5 (October 1996), p. 16.

11. This included job recognition for outstanding service and a choice of whether to work on holidays. Memo, Harold L. Boling, employee relations representative, to Carl J. Forstrum, manager, Morenci, January 3, 1986, in 5075 Shutdown Schedules, General 1985, Phelps Dodge Archive, Phoenix, AZ.

12. Interview with Priscilla Robinson, Tucson, June 6, 1988. Also letter, Darwin Aycock, secretary-treasurer of the Arizona AFL-CIO in Phoenix, to Governor Bruce Babbitt, May 30, 1986, calling on the governor to deny the 1986 operating permit to DRW as soon as possible, in file A-3-3, box 17138, EPA, Region 9 Archive, San Francisco, CA.

13. Thomas E. Sheridan, *Arizona: A History* (Tucson: University of Arizona Press, 1995), p. 170. In *Going Back to Bisbee* (Tucson: University of Arizona Press, 1992), Richard Shelton captures the ethnic ethos of these hierarchies. Michael Gonzales has written widely on the Mexican linkages, and Samuel Truett is doing important work on Hispanics and cross-border labor flows.

14. Ross R. Rice, *Carl Hayden, Builder of the American West* (Lanham, MD: University Press of America, 1994), pp. 71, 266–67. The Cerro del Pasco brief is bound in a volume of collected speeches and pamphlets entitled *The Copper Tariff*, found in the Carl Hayden papers, Arizona State University, Tucson.

15. For the linkage of cattle and copper in the 1930s, see the file in box 754, folder 17, Hayden papers.

16. U.S. Senate, Committee on Appropriations, *Hearings on the Appropriation Bill for 1937* (Washington, DC: Government Printing Office, 1938), pp. 12–13.

17. Ibid., p. 10.

18. Henry Knight Diary, May 1, 1937, p. 1003, in USDA Library, Beltsville, MD.

19. Letter, Griffin to Skinner, November 4, 1928, from Northport, in USNA, RG 97, Skinner file, entry 91, box 49.

20. Letter, F. S. Wartman from the Tucson station to R. S. Dean, December 31, 1934, in USNA, RG 70, USBM general classified files 428.1, box 791. A former USBM researcher, a Mr. Ralston headed the experimental work for United Verde.

21. Dated March 15, 1935, "The Control of Sulphur Smoke by Smelters in the United States," a seven-page typewritten report, was prepared by Dean's Metallurgical Division, in USNA, Department of State, RG 59, decimal file 711.4215, air pollution/597, box 4012, p. 2.

22. Ibid., p. 4.

23. Ibid., pp. 5–6.

24. Paul M. Ambrose, "Sulphur Dioxide in Smelter Smoke: A Review Based on Personal Observations," typewritten, [1937], p. 60, in USNA, RG 70, USBM general classified files 482.1, box 790. The leaching plant was shut down in the early 1930s. In 1936, the acid was sold at the rate of 3,600 tons per month to Inspiration and 2,700 tons to Miami. A small amount was sold to ASARCO (H. G.

Mouton, *Report* [*for the Phelps Dodge Corporation*] [New York: Phelps Dodge Corporation, 1937], pp. 15–17.

25. Ambrose, "Sulphur Dioxide in Smelter Smoke," p. 47.

26. These Mexicans had been repatriated from the United States after 1930, and their government facilitated the suit, which secured $7,500. Phelps Dodge lawyers were loath to seek U.S. help in light of the action against Trail. Dispatch of June 12, 1937, from Agua Prieta, USNA, Department of State, decimal file 711.1215, air pollution/3, box 3918.

27. Matthew Scanlon testimony, June 5, 1981, Senate Committee on Environment and Public Works, *Hearings on Clean Air Oversight*, 97th Cong., 1st sess., p. 44. Jane Kay, "PD Confirms Crop Damage Payments: Money Paid to Farmers in Mexico for 40 Years," *Arizona Daily Star* (February 8, 1986); Keoki Skinner, "Phelps Dodge Pollution Hurting Crops, Mexicans Say," *Arizona Republic* (February 9, 1986).

28. Edward F. Haase et al., Phelps Dodge Corporation, "Field Surveys of Sulphur Dioxide Injury to Crops and Assessment of Economic Damage," paper presented at the 73d meeting of the Air Pollution Control Association, Montreal, June 22–27, 1980, p. 5.

29. Notice, to "Fellow Farmers and Ranchmen," from the Sulphur Springs Valley Protective Association, McNeal, March 26, 1930. Document sent to Richard Kamp in 1979 by a valley farmer. Letter, F. S. Wartman at the Tucson station to R. S. Dean, December 31, 1934, in USNA, RG 70, USBM general classified files 428.1, box 791.

30. Letter, Mrs. Frank A. Murphy to John W. Finch, director of the USBM, from McNeal, April 30, 1935, in USNA, RG 97, USBM general classified files 482.1, box 2217.

31. Murphy to O. F. Hood, Bureau acting director, May 15, 1935, and Hayden's letter to Finch of May 20, asking if the investigation proposed by Mrs. Murphy could be arranged; and Finch's response of May 31, saying that "I do not see how this Bureau can render any service within the limitations of its Organic Act," both in USNA, RG 97, USBM general classified files 452.1, box 2217.

32. Letter, Elmer Graham, an attorney in Douglas, to Secretary Wallace, May 29, 1933, in USNA, RG 54, Records of the Bureau of Plant Industry, Records of the Bureau of Chemistry and Soils, entry 182, file 37007, smelter fumes, box 213.

33. Knight to Woods, June 6, 1933, and Acting Secretary Tugwell to Graham, June 16, 1933, USNA, RG 54, Records of the Bureau of Chemistry and Soils, entry 182, file 37007.

34. Letter, Paul R. Sipe to USDA, from Burlingame, CA, and reply from the legal office, dated August 20, 1938, in USNA, RG 17, Records of the Office of the Secretary of Agriculture, entry 17, smelter smoke (1938), box 2872.

35. Easement data from a Phelps Dodge deposition in the civil suit 70-60 filed in federal district court by Randolph H. Moses et al., dated February 20, 1971, p 32, and "Transcript of Proceedings," June 12, 1973. Easement and release HC7 408, no. 12388, in Southwest Environmental Services Archive, University of Arizona Special Collections, filed in ms. 269, box 4/16.

36. Haase et al., "Field Surveys," p. 13.

37. Civil suit number 790 was settled in the state court at Tucson in 1955. Interrogatories of PD's W. W. Little on March 24, 1971, U.S. Court of Appeals, civil suit 70-60, between Phelps Dodge Corporation, defendants, and Randolph H Moses et al., plaintiffs, Tucson, AZ, pp. 20–21, 31. Also Jane Kay, "'Old Smokey Is Not All That's Fuming in Douglas," *Arizona Daily Star* (June 14, 1981). Scanlon testimony, June 5, 1981, Senate, *Hearings on Clean Air Oversight*.

38. M. D. Lebowitz et al., "Respiratory Diseases in Tucson: Preliminary Observations," *Arizona Medicine*, Vol. 32, No. 4 (April 1975), pp 329–33, in a letter from Priscilla Robinson to Arizona Department of Health Services, January 29, 1980, in University of Arizona, records of the Southwest Environmental Service, ms 269, box 1/10. Interview with William J. Risner, a Tucson lawyer active in pollution and smelter suits during the 1970s, in Tucson, April 1, 1988. The class action suit was case 70-60, filed in federal district court in Tucson. Interview with Priscilla Robinson, Tucson, June 6, 1988.

39. "Couple Sue P-D in Crop Damage," *Arizona Daily Star* (February 10, 1982). The Mitchell case lingered on until it was settled in 1987 and accounted as a shutdown cost.

40. Isaac F. Marcosson, *Metal Magic: The Story of the American Smelting and Refining Company* (New York: Farrar Straus, 1949), p. 259. Kenneth W. Nelson, director of environmental sciences, ASARCO, "The Adequacy of the Federal Ambient Air Standards for Sulfur Dioxide in the Area of the Western Nonferrous Smelters," [1971], p. 30, in file UAP-26, Hayden Library.

41. Williams quoted in the Skinner article in the *Arizona Republic* (February 9, 1986), p. A15. Also Wayne T. Williams, "Estimates of Impacts on Agriculture in Sonora, Mexico from Copper Smelter-derived Sulfur Dioxide Pollutants," final draft, April 1986, for the Smelter Crisis Education Project, especially pp. 15–16; courtesy of Dick Kamp.

42. Haase et al., "Field Surveys," p. 14.

43. Testimony of Ralph L. Caldwell, in "Transcript of Proceedings," June 12, 1973, pp. 72–73.

44. T. M. Roberts, "Long-term Effects of Sulphur Dioxide on Crops: An Analysis of Dose-Response Relations," p. 42. He discounted the earlier work of Thomas and Katz whose fumigation cabinets were flawed. Also concluding remarks by M. W. Holdgate, p. 313, 318. Both in Sir James Beament et al., *The Ecological Effects of Deposits of Sulphur and Nitrogen Compounds*, proceedings of a Royal Society discussion meeting, September 5–9, 1983 (London: Royal Society, 1984).

45. *Phelps Dodge—A Copper Centennial, 1881–1981*, special supplement in *Pay Dirt* (Summer 1981), p. 80.

46. *Moses et al. v. Phelps Dodge*, interrogatories of February 20, 1971, pp. 27–28. Also statement by Vice President John A. Lentz before the Air Pollution Hearing Board, December 1970, p. 5, 28. Both documents courtesy of Richard Kamp. The El Paso pilot plant was shut down in April 1972 and was resumed in October 1973 on a six-month trial basis; president's letter, *Phelps Dodge Annual Report 1973*, p. 9. The precipitators, but not the scrubbers, were installed.

47. Michael Rieber, "The Economics of Copper Smelter Pollution Control: A Transnational Example," *Resources Policy* (June 1986), pp. 92–93.

48. Letter, James M. Bush to Elaine McFarland, chairman of the Arizona State Board of Health, October 20, 1971, enclosing Phelps Dodge's "Petition for Reconsideration of Arizona Sulfur Dioxide Ambient Air and Emissions Standards," in file UAP-26, Arizona State University Library. Also Phelps Dodge *Annual Report 1973*, p. 9.

49. Arizona stationary air pollution sources, 1974. Data from the Arizona Bureau of Air Quality Control in a letter from Carl Billings, manager of the Engineering Services section, to Charlotte Hopper of EPA's Region 9, from Phoenix dated November 22, 1976, in file PD Consent Decree, April 1983–December 1983, A-3-3, Air Management Division, smelter files, EPA, Region 9 Archive.

50. *Phelps Dodge—A Copper Centennial*, p. 93.

51. Interview with Professor Mike Rieber, an economist at the University of Arizona's Department of Mining and Geophysical Engineering, Tucson, June 9, 1988.

52. *Phelps Dodge—A Copper Centennial*, pp. 80–81.

53. Dick Kamp, "Modern PD Smelter Said to Be Major Polluter," *Sierra Vista Herald–Bisbee Daily Review* (January 5, 1987).

54. Letter, R. E. Johnson, superintendent of Technical Services, to A. E. Himebaugh, vice president and general manager, Phoenix, May 22, 1987, in file 109, smelters, 1987; Outokumpu [Jurö Anjala et al.], "Outokumpu Flash Smelting in Copper Metallurgy—The Latest Developments and Applications," Helsinki [1986?], in file 109, smelter, general; and Eric Groten, memorandum to files "entitled Arizona Air Pollution Control Association Meeting re New Magma Smelter," dated May 22, 1987, in file 190, general, smelter; all in Phelps Dodge Archive.

55. R. E. Johnson, Morenci smelter superintendent, and T. D. Jackson of Davo Engineers Inc., "Oxygen Sprinkle Smelting at the Morenci Smelter," paper presented to the AIME Meeting in San Francisco, November 7–9, 1983, in EPA, Region 9, file A-3-3, Air Management Division. Also, L. R. Judd, "Morenci Consent Decree and New Oxygen System," remarks to the board dated March 25, 1981, in Phelps Dodge Archive.

56. R. E. Johnson and T. D. Jackson, "Oxygen Sprinkle Smelting at the Morenci Smelter," paper presented to the AIME Meeting of the Metallurgical Society, San Francisco, November 7–9, 1983, in EPA, Region 9, file A-3-3, p. 474.

57. Larry Bowerman of the EPA to Dave Chelgren, Arizona Bureau of Air Quality Control, June 1, 1983, phone call record, in EPA, Region 9, file A-3-3.

58. Interview with Matthew P. Scanlon, former vice president at Phelps Dodge, Phoenix, November 18, 1997.

59. The quote is from Alecia Swasy's excellent analysis in the *Wall Street Journal* (November 24, 1989). Hildebrand and Mangum, *Capital and Labor in American Copper*, also cover the business turnaround, especially pp. 158–61.

60. Dun and Bradstreet Report, June 16, 1986, filed as exhibit E, in appendix of Phelps Dodge litigation report file, EPA, Region 9 Archive. The quotes are from Hildebrand and Mangum, *Capital and Labor in American Copper*, pp. 160, 195.

61. Reported in the *New York Times* (December 12, 1995), p. D1. Ilo activists alerted U.S. environmental NGOs by e-mail, and their attempts to gain standing in a U.S. district court in Texas, although rebuffed, had results. Threatened with exposure of corrupt practices and environmental mismanagement, ASARCO agreed to control the Ilo plant.

62. *Phelps Dodge—A Copper Centennial*, p. 93.

63. James T. Barrett, chief meteorologist, *Supplementary Control System, Operating Manual*, mimeo, 90 pages, from DRW Douglas, April 18, 1986, in file A-3-3, box 17133; and letter, David P. Howekamp, director of the Air Management Division, to Vice President Matthew P. Scanlon, from San Francisco, May 23, 1986, rejecting the revised SCS proposal, in box 17138; both in EPA, Region 9 Archive.

64. Letter, Frank M. Covington, director of Air and Hazardous Materials Division in San Francisco, March 26, 1976, to Suzanne Dandoy, MD, director of Arizona Department of Health Services, in EPA, file A-3-3, Air Management Division files on smelters, in box on revisions of Arizona SIP, 1981, Region 9 Archive.

65. Larry Bowerman, "Copper Smelting in EPA, Region 9," p. 1, long summary report in file Bowerman A-53, EPA, Region 9 Archive.

66. Phelps Dodge press release, "The Truth About the Douglas Smelter," [1986] in file 1040, DRW smelter, 1986, Phelps Dodge Archive; emphasis in the original.

67. Bill Epler, "P.D.'s Douglas Smelter Has Six More Months of Life—Maybe," *Southwestern Pay Dirt* (August 1986), p. 7A. Cost figures are in a letter from M. P. Scanlon, vice president for Western Operations, to Edward Reich at the EPA, from Phoenix, May 12, 1986; attachment P in EPA litigation report, in Air Management Division, EPA, Region 9 Archives.

68. Epler, "P.D.'s Douglas Smelter Has Six More Months of Life—Maybe."

69. Robinson interview, June 6, 1988.

70. Letter, N. S. Balich, controller for Western Operations, to E. J. O'Sullivan, Phoenix, October 7, 1986, file 8040, DRW 1986, and memo, "Phelps Dodge Corporation Douglas Reduction Works Shutdown Status Report, September 1987," file 999, DRW 1987, both in Phelps Dodge Archive.

71. Memo, James A. Gillen to Public Works personnel, City of Bisbee, January 15, 1987, in file 109, DRW smelter, 1987, Phelps Dodge Archive.

CHAPTER 5. MINING IS RARELY A LOCAL EVENT

1. The chapter heading is from Thomas Michael Power, *Lost Landscapes and Failed Economies: The Search for a Value of Place* (Washington, DC: Island Press, 1996), p. 119. For the Clean Air Act, see Thomas Jorling, "The Federal Law of Air Pollution Control," in *Federal Environmental Law*, ed. Erica L. Dolgin and Thomas G. P. Guilbert, pp. 1058–1103, a publication of the Environmental Law Institute (St. Paul, MN: West Publishing, 1974). The definition of public welfare in the act is quoted on p. 1084. Also, Jurgen Schmandt, Judith Clarkson, and Hilliard Roderick, eds., *Acid Rain and Friendly Neighbors: The Policy Dispute Between Canada and the United States*, rev. ed. (Durham, NC: Duke University Press, 1988), especially pp. 160–65.

2. The politics of the Clean Air Act are analyzed in Karen Ingram's classic essay, "The Political Rationality of Innovation: The Clean Air Act Amendments of 1970," in *Approaches to Controlling Air Pollution*, ed. Anne F. Friedlaender, pp. 12–56 (Cambridge, MA: MIT Press, 1978). The citation is on p. 14.

3. Ibid., p. 19.

4. Interview with David Howekamp, Air Management Division chief at Region 9, San Francisco, CA, November 10, 1994.

5. Letter, Bush to Charles Elkins, assistant administrator for Air and Radiation, Phoenix, February 26, 1986, in file 1040, DRW smelter 1986, Phelps Dodge Archive; interview with Nancy Wrona, who worked on Arizona state environmental issues in Governor Babbitt's office from 1981–1986, in Phoenix, AZ, April 1, 1988; and interview with Larry Bowerman, an air pollution specialist at EPA Region 9 ,who had been active in the smelter issue from the start, in San Francisco, CA, March 22, 1995.

6. Ingram, "The Political Rationality of Innovation," p. 22.

7. Senator Edmund Muskie, June 12, 1974, in *Congressional Record*, 93d Cong., 2d sess., vol. 120, pt. 14: 18958–59.

8. Letter, Peter Steen, president of Inspiration Consolidated Copper Company, to Congressman Eldon Rudd, Phoenix, October 21, 1981, in University of Arizona Archives, Southwest Environmental Service collection, ms. 269, box 4/18.

9. Michael Gregory testimony at EPA's 1979 Phoenix hearing, in SES records, University of Arizona.

10. Bowerman interview, March 22, 1995.

11. The pages that follow are based largely upon records of the Air Control Division in EPA's Region 9 offices in San Francisco. For charting the complex issues involving the Arizona smelters, and Douglas in particular, I found Larry Bowerman's long summary "Copper Smelting in EPA, Region 9," [1987], of great utility; in file Bowerman A-53, EPA, Region 9 Archive.

12. For example, see American Smelting and Refining Company et al., "Petition for Reconsideration and Modification of Arizona Emission and Ambient Standards for Sulfur Dioxide," [1971], in file UAP-26, Carl Hayden Library, Arizona State University. Also see Dolgin and Guilbert, *Federal Environmental Law*, p. 1085.

13. Schmandt, Clarkson, and Roderick, *Acid Rain and Friendly Neighbors*, pp. 161–62.

14. Dolgin and Guilbert, *Federal Environmental Law*, p. 1063. The case is briefly discussed in John E. Bonine and Thomas O. McGarity, *The Law of Environmental Protection*, 2d ed. (St. Paul, MN: West Publishing, 1992), pp. 387–91.

15. This synopsis follows Section III, Arizona SO_2 SIP Revisions, in Bowerman's report, "Copper Smelting in EPA, Region 9."

16. Letter, David C. Hawekins, assistant administrator for Air and Waste Management, to Governor Bruce Babbitt, April 3, 1978, in A-3-3, Air Management Division files on smelters, in box on revisions of Arizona SIP 1981, EPA, Region 9 Archive.

17. Letter, Frank M. Covington, director, Air and Hazardous Materials Division, to Suzanne Dandoy, MD, director, Arizona Department of Health Services, San Francisco, March 26, 1976, in A-3-3, Air Management Division files on smelters, EPA, Region 9 Archive. *Kennecott v. Train*, heard in the U.S. Court of Appeals, Ninth Circuit, 1975, is discussed in Bonine and McGarity, *The Law of Environmental Protection*, pp. 467–77.

18. "Petition of Phelps Dodge for Reconsideration and Repeal," [January 1978], in A-3-3, Phelps Dodge file; issues raised in the five petitions are summarized in an internal report by Larry Bowerman dated May 1, 1980; both in EPA, Region 9 Archive.

19. Interview with Nancy Wrona, April 1, 1988.

20. Letter, Babbitt to Costle, Phoenix, April 7, 1978; letter, Senator DeConcini to Costle, Washington, April 14, 1978; letter, Babbitt to Costle, May 30, 1978; and letter, Costle to Babbitt, June 12, 1978, all in EPA, Region 9 Archive.

21. The Arizona plan was developed by Jarvis Moyers and Tom Peterson of the University of Arizona. "Arizona Smelter Emission Control Plan Gains EPA Approval," *Pay Dirt* (January 1983), p. 26; Robert E. Yuhnke and Michael Oppenheimer, "Safeguarding Acid-Sensitive Waters in the Intermountain West," EDF November 26, 1984, p. 20; Bowerman interview, March 22, 1995; and Jane Kay "Arizona Lets Copper Industry Be Its Own Watch Dog," *Arizona Daily Star* (June 21, 1991).

22. Table 1, "Arizona and Nevada Smelter Sulfur Dioxide Emission Rates Based on 365-Day Year; Annual Average Emissions (tons/year), dated January 31 1985," and "Trends in Arizona Copper Smelter Sulfur Emissions, dated January 26, 1989," in Bowerman's 1987 report "Copper Smelting in EPA Region 9," in file Bowerman-A53, EPA, Region 9 Archive.

23. EDF commentary on proposed rule-making, November 30, 1981, and petition, in A-3-3, Air Management Division files on smelters, in box on revisions of Arizona SIP, 1991; also Bowerman, "Copper Smelting in EPA Region 9," pp. 2–3 in file Bowerman A-53; both in EPA, Region 9 Archive.

24. EPA Region 9, Litigation Report, 46 pages, typewritten, dated June 30, 1986, pp. 13–14; letter, Train to Goldwater, November 5, 1975, attached to the report as exhibit K; both in EPA, Region 9 Archive.

25. Bowerman, "Copper Smelting in EPA, Region 9," pp. 22–23.

26. Litigation Report, pp. 17–18. Recent stack tests showed that Douglas's two stacks emitted 1,018 pounds per hour of particulate matter compared with an allowable level of 148 pounds per hour. State data indicated that the smelter emitted 319,884 tons per year of SO_2 compared with a compliance limit of 68,657 tons per year.

27. Bowerman, "Copper Smelting in EPA, Region 9," pp. 7–8.

28. Litigation Report, pp. 23–24, and memorandum, Richard H. Mays, acting assistant administrator, to the administrator, n.d., "Enforcement Action Against Phelps-Dodge, Douglas, AZ—Penalty Issue," attachment T to the Litigation Report.

29. Litigation Report, pp. 34–35.

30. Mays memo, in Litigation Report.

31. Interview with Priscilla Robinson, Tucson, AZ, June 6, 1988.

32. Wrona interview, April 1, 1988.

33. Letter, Henderson to Region 9 Administrator Judith Ayres, Sierra Vista, AZ, April 22, 1986, in file of citizens' comments, in A-3-3, Air Division, file 703-759, box 17138, EPA, Region 9 Archive.

34. Wick to Judith Ayres, Sierra Vista, May 7, 1986, in A-3-3, Air Division, file 760-802, box 17138, EPA, Region 9 Archive; phone conversation with the author, October 29, 1997.

35. Askins to Ayres, Bisbee, May 8, 1986, in A-3-3, Air Division, file 760-802, box 17138, EPA, Region 9 Archive.

36. Andrew Gilbert III, vice president of the A. J. Gilbert Construction Company in Tucson and Bisbee, to Ayres, from Tucson, May 8, 1986, in A-3-3, Air Division, file 703-759, EPA, Region 9 Archive. Also his letter to the Bisbee Observer for April 17, 1986.

37. "Community evenly divided on smelter's benefit, harm," Douglas Daily Dispatch (February 1986).

38. August 24 in the Los Angeles Times, cited in the September 11, 1984, editorial in the Herald-Dispatch and Bisbee Review.

39. Based on the catalog description of the SES archive collection, ms. 269, at the University of Arizona; Robinson's comments to the author on her president's reports to the SES Board of Directors, December 1977–August 1987.

40. Robinson testimony before Rep. Waxman and the House Subcommittee on Health and the Environment, 97th Cong., 1st sess., December 14, 1981, p. 441, in Hearings, Clean Air Act (Part 2) (Washington, DC: Government Printing Office, 1982). After the hearings, representatives from Inspiration, Kennecott, and ASARCO sought her out. "It turned out these companies were ready to clean up, and most of the industry was ready to take on controls but didn't want PD taking unfair advantage (unfair short-term competition) if they put in controls" Robinson interview, Tucson, June 6, 1988).

41. Michael Oppenheimer, Charles B. Epstein, and Robert E. Yuhnke, "Acid Deposition, Smelter Emissions, and the Linearity Issue in the Western United States," Science, Vol. 229, No. 4716 (August 30, 1985), pp. 859–62.

42. Filed on November 28, the plaintiffs in the suit were EDF, Robinson, and two other people who lived close to the smelters. The basis of the suit was the failure of Douglas and San Manuel to comply with the SO_2 emissions limits in state

and federal regulations; the reason for the suit was to force action on the NSOs. SES Board report for September 19, 1984, SES records.

43. Chris Limberis, "Coalition Seeks Role in PD Smelter Hearing," *Arizona Daily Star* (October 17, 1985), pp. 1b–2b; memo from Bill Doe, president of SES board, to some major donors, October 29, 1985, in SES records, ms. 169, box 1/7.

44. Phone conversation with Robinson, October 20, 1997.

45. Interview with Nancy Wrona, April 1, 1988; interview with Dick Kamp in Bisbee, AZ, February 22, 1988; and interview with former governor Bruce Babbitt, June 9, 1988, in Tucson, AZ.

46. "'Acid Rain' Research in the Intermountain West, a Report by the Environmental Defense Fund of a Survey on Acid Rain Research in the Rocky Mountain Region," especially pp. 466–67, in House, *Hearings, Clean Air Act (Part 2)*, November 14, 1981.

47. Interview with Robert Yuhnke, Boulder, CO, November 13, 1987; Robinson, report to the SES Board of Directors, January 12, 1982, SES records.

48. Oppenheimer, Epstein, and Yuhnke, "Acid Deposition, Smelter Emissions, and the Linearity Issue." Dr. Edward F. Haase, "Potential Effects of Acid Precipitation in Arizona and the Southwest," conference paper presented to the Air Pollution Control Association meeting in Scottsdale, AZ, October 26–28, 1983, in SES records, ms. 269, box 5/9.

49. The argument was laid out in an EDF report published November 26, 1984, by Yuhnke and Oppenheimer, "Safeguarding Acid-sensitive Waters in the Intermountain West."

50. Yuhnke interview, November 13, 1987.

51. Memo, Yuhnke to members of the Western Caucus, Clean Air Coalition, Boulder, August 31, 1981, in SES records, ms. 269, box 4/18.

52. Failure to enforce visibility standards was the nub of its comments on Arizona's proposed SIP revision of the SO_2 standard. Letter and enclosures, Yuhnke and the Colorado Mountain Club to Administrator Anne Gorsuch, Boulder, December 15, 1981, notifying her of the intent to sue the EPA to enforce the visibility standards, EDF files, Boulder, Colorado.

53. In several hearings, EDF continued to press the western acid rain issue, as in Robert E. Yuhnke, "Testimony on Smelters and Acid Rain in the West, on behalf of EDF, the Sierra Club, and the Wyoming Outdoor Council, June 28, 1985, before the House Subcommittee on Health and the Environment," mimeo, EDF files.

54. Letter, Glyn G. Caldwell, MD, and Lee Lockie, manager, Office of Air Quality, to Yuhnke, May 22, 1986, and "Petition to Establish Operating Conditions on Smelters in Arizona to Limit SO_2 Concentrations to Prevent Endangerment of Health of Asthmatics," August 26, 1985, in A-3-3, box 17137, EPA, Region Archive.

55. Letter, Caldwell and Lockie to Yuhnke, in A-3-3, box 17137, EPA, Region 9 Archive.

56. Philip E. Enterline et al., "Report on Mortality Among Copper and Zinc Smelter Workers in the United States," Graduate School of Public Health, University of Pittsburgh, January 1986, in A-3-3, Box 17138, EPA, Region 9 Archive.

57. Letter, Yuhnke to Vale Nazar, in Office of Air, Noise, and Radiation at EPA, from Boulder February 27, 1986, and affidavit of June Hewitt, January 29, 1986, both in exhibit I in attachments to file PD Litigation Report, EPA Region Archive.

58. Wrona interview, April 1, 1988.

59. Letter, Lee Lockie, Office of Air Quality Management, ADHS, to John Kelley in Rep. Jim Kolbe's office, Phoenix, March 6, 1986, in file Air Pollution Control–DRW Operating Permit 1986, Phelps Dodge Archive.

60. Priscilla Robinson, smelter chronology for 1986; Bowerman interview, March 22, 1995; and comments by Region 9's Mark Simms to author, May 17, 1995.

61. Robinson, smelter chronology for 1986.

62. "Proposed Decision to Deny a Non-Ferrous Smelter Order to Phelps Dodge Corporation," *Federal Register*, 51:74, April 17, 1986, pp. 13085–91.

63. Videotape of the May 17, 1986, hearings on denial of the Douglas NSO through 1988, and EDF's June Hewitt tape, in EPA, Region 9 Archive.

64. This distinction is clear in a letter from Phelps Dodge lawyer James M. Bush to Congressman Raul Castro, thanking him for meeting with PD officials on border smelter matters, in which Bush was far more concerned with the acid rain issue than with the health issue; Phoenix, July 11, 1985, in file 1040, smelter 1985, Phelps Dodge Archive.

65. Interview with Richard Kamp, Bisbee, AZ, February 22, 1988.

66. DRW, "A Plan for the Modification of the Douglas Reduction Works' Supplementary Control System to Bring It into the NSO Regulations," p. 6, in A-3-3, box 17138, EPA, Region 9 Archive. Donald W. Moon, "The Diurnal Variation of Ambient SO$_2$ Measurements in the Vicinity of Arizona Copper Smelters" (master's thesis, Arizona State University, 1974), pp. 17, 24.

67. Richard Kamp, "Comments on Smelter Regulations," sent to Senator Stafford's Committee on Environment and Public Works, June 17, 1981, in box 4, file 18, SES Archive.

68. Robinson, chronology for 1983, and June 14 report to the SES board. The hearing held in Tucson on May 20, 1983, examined "U.S. Assistance to Foreign and Domestic Industries and Environmental Standards," House Committee on Interior, Subcommittee on Mining, 98th Cong., serial no. 98-21 (Washington, DC: Government Printing Office, 1984).

69. Kamp, "Request for 12-Month Data Assessment and Public Education Funding: Mexican-American Smelter Pollution Crisis," mimeo, Naco, AZ, September 30, 1983, courtesy of Dick Kamp.

70. Kamp interview, February 22, 1988. Jane Kay, "Blue Skies Turn Gray in the 'Triangle,'" *Arizona Daily Star* (August 29, 1983), p. 13E.

71. Richard Kamp, "Time Short to Stop Impending Smelter Menace," *Herald-Dispatch and Bisbee Review* (October 23, 1983).

72. Yuhnke interview, November 13, 1987.

CHAPTER 6. THE GRAY TRIANGLE

1. *Phelps Dodge—A Copper Centennial, 1881–1981*, special supplement in *Pay Dirt* (Summer 1981), pp. 111–14, 123–24; H. G. Moulton, *Report* [for the Phelps Dodge Corporation] (New York: Phelps Dodge Corporation, 1937), pp. 96–99; Isaac F. Marcosson, *Metal Magic: The Story of the American Smelting and Refining Company* (New York: Farrar, Straus, 1949), chaps. 10 and 11.

2. Skinner, "Phelps Dodge Pollution Hurting Crops, Mexicans Say," *Arizona Republic* (February 9, 1986), pp. A14–15.

3. Smelter Crisis Education Project, flyer, June 6, 1985, and David Chelgren, "Smelter Report," July 1985, both in University of Arizona, SES records, ms. 269,

Box 5/12. Jane Kay reported on the Bureau's relaxed enforcement policies in "Arizona Lets Copper Industry Be Its Own Watchdog," *Arizona Daily Star* (June 21, 1981).

4. Stephen P. Mumme, "Dependence and Interdependence in Hazardous Waste Management Along the U.S.-Mexico Border," *Policy Studies Journal*, Vol. 14, No. 1 (September 1985), p. 161, and letter, Robert J. McKinsey at EXIM Bank to Priscilla Robinson, Washington, D.C., June 6, 1983, SES records, ms. 269, box 5/5.

5. Statement before the Health and Environment Subcommittee of the House Committee on Energy and Commerce, Albuquerque, NM, June 28, 1985; hearing documents courtesy of EDF.

6. Ibid.

7. Priscilla Robinson, smelter chronology for 1984 and 1985.

8. Stephen P. Mumme, "State and Local Influence in Transboundary Environmental Policy Making Along the U.S. Mexico Border: The Case of Air Quality Management," *Journal of Borderlands Studies*, Vol. 2, No. 1 (1986), pp. 2–6.

9. Interview with Alberto Székely and Roberta Lajous, Mexico City, December 7, 1989, and interview with Cliff Metzger in San Diego, CA, April 7, 1995.

10. Mumme, "State and Local Influence," p. 2.

11. Metzger interview, April 7, 1995.

12. Alejandra Lajous Vargas, coordinator, *Crónica del Sexenio 1982–1988* (Mexico: Presidencia de la República, 1985–1988), 2:39 and 6:149–50.

13. Székely and Lajous interview, December 7, 1989.

14. Michael Rieber, "The Economics of Copper Smelter Pollution Control: A Transnational Example," *Resources Policy* (June 1986), pp. 101–2.

15. Stephen P. Mumme, C. Richard Bath, and Valerie J. Assetto, "Political Development and Environmental Policy in Mexico," *Latin American Research Review*, Vol. 23, No. 1 (Winter 1988), p. 24.

16. Testimony of Dr. Philip Landrigan, chief, Environmental Hazards Activity, in hearing held in Hopewell, Virginia, on January 30, 1976, by the Subcommittee on Manpower, Compensation, and Health and Safety, U.S. Senate, 94th Cong., 2d sess., 1976, pt. 2: 348–50.

17. Metzger interview, April 7, 1995.

18. Interview with former governor Bruce Babbitt, June 8, 1988, Tucson, AZ.

19. Letter, Babbitt to Lee Thomas, Phoenix, March 27, 1986, in SES records.

20. Székely and Lajous interview, December 7, 1989.

21. SEDUE, *Informe de Labores, 1985–1986* (Mexico, n.d.), pp. 57–58.

22. Quoted by Mumme, "State and Local Influence," p. 9, who gives an excellent chronology of events from the U.S. side.

23. Ibid., pp. 6–7.

24. Székely and Lajous interview, December 7, 1989.

25. Convenio, June 1, 1986, mf 100010000/55, Centro de Documentación de la Gestión Gubernamental, Coyoacán Archive, Mexico City.

26. Dick Kamp, "Mexican Government Steps in to Run Nacozari Mine, Smelter," *Bisbee Review/Sierra Vista Herald* (July 24, 1988), p. 7A.

27. Vargas, *Cronica del Sexenio*, 4:421.

28. Mexican production statistics from Subsecretaría de Minas e Industria Basica, "Informe de las funciones realizadas por las dirección general de minas correspondientes a los años de 1983 a 1987," mimeo, Coyoacán Archive. Also Rieber, "The Economics of Copper Smelter Pollution Control," p. 97.

29. Smelter statistics from International Wrought Copper Council, *Survey of Capacities of Copper Mines, Smelters . . . ,* pt. 1 (London, summer, 1987), p. 33.

Armando Durazo, "Copper Needs Across Border Collide," *Tucson Citizen* (February 9, 1983), pp. 1C, 4C.

30. Informe de actividades de INE, July, 1984, on the dumping of nuclear wastes in Juárez in June, mf 05030200/2, Coyocán Archive, Mexico City, and SEDUE, *Informe de Labores, 1985–1986* (Mexico City, n.d), p. 58.

31. Székely and Lajous interview, December 7, 1989.

32. Babbitt interview, June 8, 1988; Richard Kamp interview, Bisbee, AZ, February 22, 1988; and Mumme, "State and Local Influence," p. 6.

33. Interview with Robert Yuhnke, November 13, 1987, Boulder, CO.

34. Organized by Helen Ingram and Robert Varady, the U.S.-Mexico Conference on Border Smelter Emissions (February 6–7, 1986) was one of the first attempts by academic specialists to examine border environmental issues within the context of a rapidly intensifying bilateral relationship that demanded cooperative approaches.

35. Interview with David Howekamp, November 10, 1994, San Francisco, CA, and Metzger interview, April 7, 1995.

36. The growing independence of Mexican NGOs is discussed in the introduction to Richard Kiy and John D. Wirth, eds., *Environmental Management on North America's Borders* (College Station: Texas A&M University Press, 1998).

37. Dick Kamp, "Border Pollution May Look Different on the Other Side," interview with Roberto Sánchez, *Bisbee Review/Sierra Vista Herald* (April 24, 1988).

38. Albert Utton, memorandum to U.S.–Mexico Working Group on the research agenda, March 28, 1985; Székely's remarks were given at the Puerto Vallarta workshop in December 1984; both in SES records, ms. 269, 4/18.

39. Richard Kamp, "Douglas Smelter's Future Contribution to Border 'Triangle' Subject of Debate," *Herald-Dispatch and Bisbee Review* (May 6, 1984).

40. Ibid.

41. [James T. Barrett, chief meteorologist], *Supplementary Control System, Operating Manual,* mimeo, 90 pages from DRW, in file A-3-3, box 17133, EPA, Region 9 Archive.

42. These negotiations are outlined in "Proposed Decision to Deny a Non-Ferrous Smelter Order to Phelps Dodge Corporation, Douglas, AZ; Section 119 of the Clean Air Act," in the *Federal Register,* Vol. 51, No. 74 (April 17, 1986), pp. 13085–89.

43. Matthew P. Scanlon interview, Phoenix, AZ, November 18, 1997.

44. Letter, John F. Boland, environmental lawyer, to Senator Dennis DeConcini, November 19, 1982, urging him not to support a bill advocated by the steelworkers to impose the acid plant, in file 1040, DRW smelter, 1986, Phelps Dodge Archive.

45. Letter, Stan Turley, president of the Senate, et al., to George P. Shulta and Lee M. Thomas, Phoenix, July 8, 1985, in file 1040, DRW smelter, 1986, Phelps Dodge Archive.

46. This was the company's view; letter, James M. Bush to Charles L. Elkins, assistant administrator for Air and Radiation, Phoenix, February 26, 1986, in file 1040, DRW smelter, 1986, Phelps Dodge Archive. The environmentalists thought that Babbitt, caving in to the legislature, was backing away from what they thought was an earlier commitment to deny the state operating permit; Mumme, "State and Local Influence," p. 9. Kamp saw him as a Machiavellian politician, but an environmentalist at heart; Kamp interview, February 22, 1988.

47. Bush letter, February 26, 1986, in file 1040, DRW smelter, 1986, Phelps Dodge Archive.

48. Babbitt to Thomas, March 27, 1986, same source as note 46.

49. *Federal Register* for April 17, 1986, pp. 1389–1409, which referenced Babbitt's letters to Lee Thomas of August 21, 1985, and February 11, 1986.

50. "The Truth About the Douglas Smelter," press release from the office of Leonard R. Judd, senior vice president, n.d. [1986], in file 1040, DRW smelter, 1986, Phelps Dodge Archive.

51. This attitude is changing, as I discovered firsthand in 1988 while serving on the Advisory Committee for the CEC's project to protect the San Pedro River riparian zone, a major bird flyway between Mexico, the United States, and Canada. Arizonans on the border realize that safeguarding the river from overuse can only be assured through a bilateral, cooperative program with Mexico.

52. Dick Kamp, "Nacozari Smelter: Pollution Control and Human Realities," *Border Geology Project Newsletter* (May 5,1988).

CHAPTER 7. CONTINENTAL PATHWAYS

1. Mira Wilkins, *The Emergence of Multinational Enterprise: American Business Abroad from the Colonial Era to 1914* (Cambridge, MA: Harvard University Press, 1970), pp. 115–16, 124–25.

2. James Robert Allum, "Smoke Across the Border: The Environmental Politics of the Trail Smelter Investigation" (Ph.D. diss., Queens University, Kingston, ONT, 1995), pp. 32–37, 346.

3. Theodore C. Scheffer and George C. Hedgcock, "Injury to Northwestern Forest Trees by Sulfur Dioxide from Smelters," USDA, Forest Service, *Technical Bulletin* no. 1117 (June 1955), p. 43.

4. R. C. Crowe, ed., preface to *Trail Smelter Question* (Trail, BC: Consolidated Mining and Smelting Company, 1939).

5. John E. Read, "The Trail Smelter Dispute," *Canadian Yearbook of International Law*, Vol. 1 (1963), p. 227.

6. Jamie Alley, "The British Columbia–Washington Environmental Cooperation Council: An Evolving Model of Canada-U.S. Interjurisdictional Cooperation," in *Environmental Management on North America's Borders*, Richard Kiy and John D. Wirth, pp. 53–71 (College Station: Texas A&M University Press, 1998).

7. British Columbia–Washington Environmental Cooperation Council. "Northport Presentation," October 16, 1995, Victoria, BC. Also Washington State Department of Ecology, "Northport Chronology [1992–1997]," unpublished fact sheet, and J. VanEenwyck, "Approaches to Community Concerns: Applied Public Health [in Northport]," *Public Health*, Vol. 111 (1997), pp. 405–10.

8. Terry Glavin, "Toxic Chemical Worries Straddle the Border," *Vancouver Sun* (November 19, 1991), p. B13.

9. "Despite the leaching of metals from Cominco slag and the discharge of metals contaminated waste water from Cominco Ltd., the concentration of total metals, as measured in Lake Roosevelt water samples collected near Northport, did not exceed existing U.S. Environmental Protection Agency drinking water standards or ecological criteria" (Glen Patrick, "Cominco Slag in Lake Roosevelt; Review of Current Data," December 1993, appendix D, p. 4, in *Air Monitoring Data and Evaluation of Health Concerns in Areas of Northeast Tri-County* [Olympia, WA: Department of Health, April 1994]).

10. British Columbia–Washington Environmental Cooperation Council, *Fourth Annual Report to the Governor and the Premier* (Victoria, BC, and Olympia, WA, 1996).

11. International Joint Commission, *The IJC and the 21st Century* (n.p., 1997), pp. 27–37. Commission for Environmental Cooperation, *Continental Pollutant Pathways: An Agenda for Cooperation to Address Long-Range Transport of Air Pollution in North America* (Montreal: CEC, 1997).

12. Randy Lee Loftis, "Panel Rejects Nuclear Site in West Texas," *Dallas Morning News* (October 23, 1998), www.texasalmanac.com. Jorge Olmedo y Filiberto Cruz, "Positiva respuesta a gestiones de México, declaran SRE y SEMARNAP," *El Sol de México* (October 23, 1998). José Manual Nava, "Beneplácto de ambientalists por el fallo contra el projecto," *Excelsior* (October 25, 1998).

13. Bill Powers and Richard Kamp, "Mexican Smelters in the Border Region," pp. 25–30, in *Technical Basis for Appendices to Annex IV of the La Paz Agreement*, EPA-456/R-97-xxx, September 1997.

14. For a succinct analysis of the issues, see Mary Kelly, "Carbón I/II: An Unresolved Binational Challenge," pp. 189–207, and Peter Emerson et al., "Managing Air Quality in the Paso del Norte Region," pp. 125–163, both in King and Wirth, *Environmental Management on North America's Borders*.

15. CEC, *Continental Pollutant Pathways*, pp. 1, 31.

16. Victor Lichtinger, former director of the CEC, in preface, *Continental Pollutant Pathways*.

17. This summary of current air pollution issues is adopted from the section on air, pp. 49–58, in Commission for Environmental Cooperation, "On Track? Sustainability and the State of the North American Environment," draft, November 4, 1998.

BIBLIOGRAPHY

UNPUBLISHED DOCUMENTS

Government Archives

Canada, Department of External Affairs, in National Archives (CNA)Trail Smelter Arbitration Case, RG 25.
Papers of Secretary of State O. D. Skelton, microfilm reel T-1757, RG 25.
Canada, National Research Council, in CNA
"Trail Smelter Investigation," Vols. 90–91, August 1938–January 1949, RG 97.
Associate Committee on Trail Smelter Smoke, minutes, uncataloged.
Chemistry, Box 5, including Morris Katz's Chemistry Progress Reports, September 1938–April 1940, uncataloged.
Mexico, Centro de Documentación de la Gestion Gobernamental, Miguel de la Madrid Presidential Papers, Coyoacán Archive, Mexico City, DF.
State of Washington, papers relating to Trail smelter, including Governor Hartley papers, Washington State Archives, Olympia, WA.
United States, Department of Agriculture, in National Archives (USNA)
Bureau of Chemistry and Soils, 1927–1936, RG 54, 1936–1942, RG 97, including Bureau Chief Henry Knight's correspondence and office files of W. W. Skinner, 1927–1941, RG 97.
Bureau of Plant Industry, Soils, and Agricultural Engineering, RG 54, including Records of the Office of the Chief Chemist, Stewart W. Griffin, 1928–1948, and Bureau Chief's correspondence, and records relating to the Trail Smelter Fumes Investigation, 1927–1934.
Office of the Secretary, RG 16.
United States, Department of Agriculture, Henry Knight Diary, 1929–1942, 2 volumes, typewritten, in USDA Library, Beltsville, MD.
United States, Department of Interior, in USNA Bureau of Mines, general classified files, smelter fumes, RG 70.
United States, Department of State, in USNA
Air pollution, RG 59, decimal file 711.4215.
Trail Smelter Arbitration, including reports on scientific investigations, 1926–1940, RG 76.
Agua Prieta Consular reports, 1930–1939, decimal file 711.1215.
United States Environmental Protection Agency, Region 9, Air and Toxics Division, records pertaining to environmental compliance of Arizona smelters, San Francisco, CA.
United States Office of Personnel Management, personnel files of USDA scientists, St. Louis, MO.

Corporate Archives

Consolidated Mining and Smelting Company (COMINCO)
Extensive holdings of reports, legal briefs, hearing documents, and other doc-

uments related to the Trail smelter, 1917–1941, add. mss. 2500, British Columbia Archives, Victoria, BC.
Papers relating to the Trail smelter, 1923–1941, Vancouver, BC.
Phelps Dodge Corporation, papers relating to environmental compliance and
shutdown of the Douglas Reduction Works, 1980s, Phoenix, AZ.

Other Archives

Arizona State University, Carl Hayden Library including Senator Hayden's pa-
pers, Tempe, AZ.
Border Ecology Project, records pertaining to the Gray Triangle, 1979–1988, Bis-
bee, AZ.
Environmental Defense Fund, documents, reports, and memoranda relating to
Douglas and western acid rain, Boulder, CO.
Stanford University, archives pertaining to Professor Robert E. Swain and Board
of Trustees, and Hoover Institution documents on the Trail Smelter Arbitral
Commission, 1936–1941, including *Hearings*, Stanford, CA.
Stevens County Historical Society, newspaper collection, Lieutenant Governor Lon
Johnson's unpublished "Memoirs," and miscellaneous documents, Colville, WA.
University of Arizona, records of the Southwest Environmental Service,
1978–1986, Tucson, AZ.

PUBLISHED DOCUMENTS AND REPORTS

British Columbia–Washington Environmental Cooperation Council. *Fourth Annual
Report to the Governor and the Premier.* Victoria, BC, and Olympia, WA, 1996.
Commission for Environmental Cooperation. *Continental Pollutant Pathways: An
Agenda for Cooperation to Address Long-Range Transport of Air Pollution in North
America.* Montreal, QUE: CEC, 1997.
———. "On Track? Sustainability and the State of the North American Environ-
ment." Draft, November 4, 1998.
———. *Report of the CEC Upper San Pedro River Initiative Advisory Panel.* Montreal,
QUE: CEC, November, 1998. November 1998.
Crowe, R. C., ed. *Trail Smelter Question* (in two parts). Trail, BC: Consolidated
Mining and Smelting Company, 1939.
Dean, Reginald, and Robert E. Swain. "Report Submitted to the Trail Smelter Ar-
bitral Tribunal." In U.S. Bureau of Mines, *Bulletin* 453. Washington, DC: Gov-
ernment Printing Office, 1944. The original, unaltered "Report of the Technical
Consultants to the Trail Smelter Arbitral Tribunal; Meteorological Observations
Near Trail, B.C., 1938–1940," in typescript, may be found in the BC Provincial
Archives, add. mss. 2500, and in Swain's papers at the Hoover Institute.
[Endersby, Alfred]. *An Appeal from the Court of Appeal for British Columbia Between
Consolidated Mining and Smelting Company and Alfred Endersby Plaintiff.* Victoria,
BC, 1919, in BC Archives, add. mss. 2500.
Holmes, A., et al., "Report of the Selby Smelter Smoke Commission." In U.S.
Bureau of Mines, *Bulletin* 98. Washington, DC: Government Printing Office,
1915.
[Howes, E. A., and F. G. Miller]. *The Deans' Report: Final Report to the I.J.C. by Dean
E. A. Howes and Dean F. G. Miller.* Published as appendix A1 of the *Trail Smelter
Question.* Ottawa, ONT: Government Printing Office, 1936.

International Joint Commission. Trail smelter reference. *Brief Submitted on Behalf of His Majesty's Government in Canada.* Ottawa, ONT: Government Printing Office, 1930.

——. *The IJC and the 21st Century.* N.p., 1997.

[Katz, Morris, and F. E. Lathe et al.] *Effect of Sulfur Dioxide on Vegetation.* Prepared for the Associate Committee on Trail Smelter Smoke of the National Research Council of Canada. Ottawa, ONT: National Research Council, 1939.

Mexico, Presidencia. Lajous Vargas, Alejandra, coordinadora. *Crónica del Sexenio 1982–1988.* Vols. 2 and 6. Mexico, DF: Presidencia de la República, 1985–1988.

Mexico, Secretaría de Desarollo Urbana y Ecología (SEDUE). *Informe de Labores, 1985–1986.* Mexico, DF: n.d.

Moulton, H. G. *Report* [for the Phelps Dodge Corporation]. New York: Phelps Dodge Corporation, 1937.

National Research Council. *Twenty-first Annual Report, 1937–1938.* Ottawa, ONT: NRC, 1938.

Phelps Dodge Corporation. *Annual Reports,* 1970–1990.

Scheffer, Theodore C., and George C. Hedgcock. "Injury to Northwestern Forest Trees by Sulfur Dioxide from Smelters." In USDA, Forest Service, *Technical Bulletin* no. 1117 (June 1955), pp. 1–47.

State of Washington, Department of Ecology.
  Johnson, Art, et al. *An Assessment of Metals Contamination in Lake Roosevelt.* Water Body no. WA-CR-9010, revised December 1989. Olympia, WA: Ecology Department, June 1988.
  ——. *Spatial Trends in TCDD/TCDF Concentrations in Sediment and Bottom Fish Collected in Lake Roosevelt (Columbia River).* Water Body no. WA-CR-9010. Olympia, WA: Ecology Department, June 1991.
  "Northport Air Quality Study" [1996?].
  "Northport Chrolonogy" [1997].

State of Washington, Department of Health. *Air Monitoring Data and Evaluation of Health Concerns in Areas of Northeast Tri-County.* Olympia, WA: Health Department, April 1984.

Trail Smelter Arbitral Tribunal. *Decision of the Tribunal, Reported March 11, 1941.* Washington, DC: Government Printing Office, 1941.

Trail Smelter Arbitral Tribunal. *Trail Smelter Question.* Documents Series. Ottawa, ONT: Government Printing Office, 1937.

Trail Smelter Arbitral Tribunal. *Trail Smelter Question, Hearings.* Washington, DC: Government Printing Office, 1941.

United States Government.
  Congressional Record. Senate. June 12, 1974, pp. 18964–71.
  Environmental Protection Agency. Bill Powers and Dick Kamp. *Technical Basis for Appendices to Annex IV of the La Paz Agreement.* EPA-456/R-97-xxx, September, 1997.
  Federal Register. April 17, 1986, pp. 13085–89.
  House of Representatives.
  Committee on Appropriations. *Hearings* and *Reports.* 1930–1937.
  Subcommittee on Health and the Environment. *Hearings, Clean Air Act (Part 2).* 97th Congress, first session, December 14, 1981. Washington, DC: Government Printing Office, 1982.
  Subcommittee on Manpower, Compensation, and Health and Safety. *Hearings (Part 2).* 94th Congress, second session, January 30, 1976. Washington, DC: Government Printing Office, 1977.

Senate.
Committee on Appropriations. *Hearings*. 1932–1937. Committee on Environment and Public Works. *Hearings on Clean Air Oversight*. 97th Congress, June
5, 1981. Washington, DC: Government Printing Office, 1981.

UNPUBLISHED DOCUMENTS, BOOKS, AND ARTICLES

Allum, James Robert. "Smoke Across the Border: The Environmental Politics of
the Trail Smelter Investigation." Ph.D. dissertation, Queens University, Kingston, ONT, 1995.
Bowerman, Larry. "Copper Smelting in EPA, Region 9." Long summary report
[1987], in file Bowerman A-53, EPA, Region 9 Archive.
Cioc, Mark. "The Trail Smelter Dispute: International Arbitration and Trans-
Boundary Air Pollution in the Columbia Basin, 1927–1941." Paper presented
to the American Society for Environmental History, Pittsburgh, Pa., March
1993.
Haase, Edward F. "Potential Effects of Acid Precipitation in Arizona and the
Southwest." Paper presented to the Air Pollution Control Association, Scotts-
dale, AZ, October 26–28, 1983.
Haase, Edward F., et al., Phelps Dodge Corporation. "Field Surveys of Sulphur
Dioxide Injury to Crops and Assessment of Economic Damage." Paper pre-
sented at the 73d meeting of the Air Pollution Control Association, Montreal,
June 22–27, 1980.
Hall, Ridgway H. "The Lost Legacy of the Trail Smelter." Undergraduate honors
thesis, History Department, Yale University, 1989.
Johnson, Arthur Beers. "A Study of the Action of Sulphur Dioxide at Low Con-
centrations on the Wheat Plant, with Particular Reference to the Question of
'Invisible Injury.'" Ph.D. dissertation, Department of Chemistry, Stanford Uni-
versity, 1932.
MacMillan, Donald. "A History of the Struggle to Abate Air Pollution from Cop-
per Smelters of the Far West, 1885–1933." Ph.D. dissertation, History Department, University of Montana, 1973.
Metcalfe, K. A. "A History of Sulfur Dioxide Abatement Processes at COMINCO's
Trail Smelter." Paper Presented to Globe '90 Conference, Vancouver, BC, March
22, 1990.
Moon, Donald W. "The Diurnal Variation of Ambient SO_2 Measurements in the
Vicinity of Arizona Copper Smelters." Master's thesis, Arizona State University, 1974.
Nelson, Kenneth W., director of environmental sciences, ASARCO. "The Adequacy of the Federal Ambient Air Standards for Sulfur Dioxide in the Area of
the Western Nonferrous Smelters." Typewritten report, [1971], in file UAP-26,
Hayden Library, Arizona State University.
Quivik, Frederic L. "Conflict in the Science of Environmental Impact: The Anaconda Smelter Smoke Cases, 1902–1910." Paper presented to the American Society for Environmental History, Las Vegas, NV, March 1995.
———. "Smoke and Tailings: An Environmental History of Copper Smelting Technologies in Montana, 1880–1938." Ph.D. dissertation, University of Pennsylvania, Philadelphia, 1998.
Snyder, Lynne Page. "United States Steel, Pittsburgh Farmers, and the Politics of
Environmental Science: Air Pollution Near Denora, Pennsylvania, Before the

Smog, 1915–1942." Paper presented at California University of Pennsylvania, California, PA, April 17, 1993.

Stevens County Superior Court, Civil Court, cases 6542 and 6600, microfilm 146, 147. Colville, WA.

Trail Smelter Arbitral Tribunal. "Decision . . . of April 16, 1938." Typewritten, in CNA, RG 25.

United States Court of Appeals. Civil Suit 70-60, between Phelps Dodge Corporation, defendants, and Randolph H. Moses et al., plaintiffs. Tucson, AZ, 1971.

Warren, Christian. "The Silenced Epidemic: A Social History of Lead Poisoning in the United States Since 1900." Ph.D. dissertation, Brandeis University, Waltham, MA, 1997.

Whittaker, Lance H. "All is Not Gold." History of the Consolidated Mining and Smelting Company, Trail, BC, 1945. Typewritten, in BC Provincial Archives, add. mss. 2500.

Williams, Wayne T. "Estimates of Impacts on Agriculture in Sonora, Mexico, from Copper Smelter–Derived Sulfur Dioxide Pollutants." Paper for the Smelter Crisis Education Project, April 1986, Border Ecology Project.

PUBLISHED BOOKS AND ARTICLES

Aguilar Camín, Héctor. La frontera nómada: Sonora y la Revolución Mexicana. Mexico, DF: Cal y arena, 1997.

———. "Arizona Smelter Emission Control Plan Gains EPA Approval." Pay Dirt (January 1983), p. 26.

Barman, Jean. The West Beyond the West: A History of British Columbia. Toronto, ONT: University of Toronto Press, 1991.

Bernstein, Marvin D. The Mexican Mining Industry, 1890–1950. Albany: State University of New York, 1964.

Blank, Stephen, et al. "North American Business Integration." Business Quarterly, Vol. 58, No. 3 (1994), pp. 55–61.

Blank, Stephen, and Jerry Haar. Making NAFTA Work: U.S. Firms and the New North American Business Environment. Miami, FL: North-South Center Press, 1998.

Bonine, John E., and Thomas O. McGarity. The Law of Environmental Protection. 2d ed. Saint Paul, MN: West Publishing, 1992.

Byrkit, James W. Forging the Copper Collar: Arizona's Labor-Management War, 1901–1921. Tucson: University of Arizona Press, 1982.

Caldwell, Lynton Keith. "Binational Responsibilities for a Shared Environment." In Canada and the U.S.: Enduring Friendship, Persistent Stress, ed. Charles F. Duran and John H. Sigler. New York: Prentice Hall, 1985.

———. "Canada and the American Institute." Engineering and Mining Journal, Vol. 115, No. 18 (May 5, 1923), p. 787.

Carroll, John E. Environmental Diplomacy: An Examination and a Perspective of Canadian-U.S. Transboundary Environmental Relations. Ann Arbor: University of Michigan Press, 1983.

———. "Transboundary Air Quality Relations: The Canada–United States Experience." Canadian-American Public Policy, No. 2 (July 1990), pp. 1–23.

Chaput, Donald, and Kett Kennedy. The Man from ASARCO: A Life and Times of Julius Kruttschnitt. Parkville, Australia: Australian Mineral Heritage Trust and Australasian Institute of Mining and Metallurgy, 1992.

Cleland, Robert Glass. *A History of Phelps Dodge, 1834–1950.* New York: Alfred A. Knopf, 1952.

Dässler, H.-G., and S. Börtitz. *Air Pollution and Its Influence on Vegetation.* Dordrecht, Netherlands: Junk Publishers, 1988.

Dietrich, William. *Northwest Passage: The Great Columbia River.* New York: Simon and Schuster, 1995.

Dinwoodie, D. H. "The Politics of International Pollution Control: The Trail Smelter Case," *International Journal,* Vol. 27, No. 2 (Spring 1972), pp. 218–35.

Dolgin, Erica L., and Thomas G. P. Guilbert, eds. *Federal Environmental Law* (a publication of the Environmental Law Institute). St. Paul, MN: West Publishing, 1974.

Dominick, Raymond H., III. *The Environmental Movement in Germany: Prophets and Pioneers, 1871–1971.* Bloomington: Indiana University Press, 1992.

Eggleston, Wilfred. *National Research in Canada: The NRC, 1916–66.* Toronto, ONT: Irwin Clarke, 1978.

Epler, Bill. "P.D's Douglas Smelter has Six More Months of Life—Maybe." *Southwestern Pay Dirt* (August 1986), pp. 7–8A.

———. "Fume Troubles in the Offing." Editorial. *Chemical and Metallurgical Engineering,* Vol. 44, No. 4 (April 1937), p. 179.

Gonzales, Michael J. "United States Copper Companies, the Mine Workers' Movement, and the Mexican Revolution, 1910–1920." *Hispanic American Historical Review,* Vol. 76, No. 3 (August 1996), pp. 503–34.

———. "United States Copper Companies, the State, and Labor Conflict in Mexico, 1900–1910." *Journal of Latin American Studies,* Vol. 26, Part 3 (October 1994), pp. 651–81.

Griffin, S. W., and W. W. Skinner. "Small Amounts of Sulfur Dioxide in the Atmosphere: An Improved Method for the Determination of Sulfur Dioxide when Present in Very Low Concentration in the Air." *Industrial and Engineering Chemistry,* Vol. 24 (1932), pp. 862–67.

Hadley, Diana. "Border Boom Town—Douglas, Arizona, 1900–1920." *Cochise Quarterly,* Vol. 17, No. 3 (Fall 1987), pp. 3–47.

Halliday, C. F. "A Historical Review of Atmospheric Pollution." World Health Organization, monograph series no. 46, *Air Pollution.* Geneva: WHO, 1961.

Harden, Blaine. *A River Lost: The Life and Death of the Columbia.* New York: Norton, 1996.

Hart, Michael. *Decision at Midnight: Inside the Canada-U.S. Free-Trade Negotiations.* Vancouver: University of British Columbia Press, 1994.

Heck, Walter W. "Assessment of Crop Losses from Air Pollutants in the United States." In *Air Pollution's Toll on Forests and Crops,* ed. James J. MacKenzie and Mohamed T. El-Ashry. New Haven, CT: Yale University Press, 1989.

Heyman, Josiah McC. "In the Shadow of the Smokestacks: Labor and Environmental Conflict in a Company-Dominated Town." In *Articulating Hidden Histories: Exploring the Influence of Eric R. Wolf,* ed. Jane Schneider and Rayna Rapp, pp. 156–74. Berkeley: University of California Press, 1995.

———. *Life and Labor on the Border: Working People of Northeastern Sonora, Mexico, 1886–1986.* Tucson: University of Arizona Press, 1991.

———. "The Oral History of the Mexican-American Community of Douglas, Arizona, 1901–1942." *Journal of the Southwest,* Vol. 35, No. 2 (Summer 1993), pp. 186–205.

Hicks, John D. *Agricultural Discontents.* Madison: University of Wisconsin Press, 1951.

Hildebrand, George H., and Garth L. Mangum. *Capital and Labor in American Copper, 1845–1990: Linkages Between Product and Labor Markets.* Cambridge, MA: Harvard University Press, 1992.

Hill, George R., and M. D. Thomas. "Influence of Leaf Destruction by Sulphur Dioxide and by Clipping on Yield of Alfalfa." *Plant Physiology,* Vol. 8 (1933), pp. 223–45.

Hutchinson, Eric. *The Department of Chemistry, Stanford University, 1891–1976.* Stanford, CA: Stanford University Press, 1977.

Hutchinson, Thomas C. "The Ecological Consequences of Acid Discharges from Industrial Smelters." In *Acid Precipitation: Effects on Ecological Systems,* ed. Frank M. D'Intri. Ann Arbor, MI: Science Publishers, 1982.

Ingram, Karen. "The Political Rationality of Innovation: The Clean Air Act Amendments of 1970." In *Approaches to Controlling Air Pollution,* ed. Anne F. Friedlander, pp. 12–56. Cambridge, MA: MIT Press, 1978.

International Wrought Copper Council. *Survey of Capacities of Copper Mines, Smelters. . . .* Part 1 (London: IWCC, Summer 1987).

Katz, Morris. "Some Aspects of the Physical and Chemical Nature of Air Pollution." World Health Organization, monograph series no. 46, *Air Pollution.* Geneva: WHO, 1961.

———. "Sulfur Dioxide in the Atmosphere and Its Relation to Plant Life." *Industrial and Engineering Chemistry,* Vol. 41, No. 11 (November 1949), pp. 2450–66.

Kiy, Richard, and John D. Wirth, eds. *Environmental Management on North America's Borders.* College Station: Texas A&M University Press, 1998.

Kluge, Thomas, and Engelbert Schramm. "Von Himmel hoch, Eine Geschichte der TA Luft." *Kursbuch,* Vol. 96 (1989), pp. 91–109.

Krupa, Sagar V. *Air Pollution, People, and Plants.* St. Paul, MN: American Phytopathological Society, 1997.

Lamborn, John E., and Charles S. Peterson. "The Substance of the Land: Agriculture v. Industry in the Smelter Cases of 1904 and 1906." *Utah Historical Quarterly,* Vol. 53, No. 4 (Fall 1985), pp. 308–25.

Lebowitz, M. D., et al. "Respiratory Diseases in Tucson: Preliminary Observations." *Arizona Medicine,* Vol. 32, No. 4 (April 1975), pp. 329–33.

Malone, Patrick. "Butte: Cultural Treasure in a Mining Town." *Montana, the Magazine of Western History,* Vol. 47, No. 4 (Winter 1997), pp. 58–67.

Marcosson, Isaac F. *Metal Magic: The Story of the American Smelting and Refining Company.* New York: Farrar, Straus, 1949.

Mickelson, Karin. "Rereading *Trail Smelter.*" *Canadian Yearbook of International Law,* Vol. 31, Tome 31 (1993), pp. 219–33.

Mouat, Jeremy. *Roaring Days: Rossland's Mines and the History of British Columbia.* Vancouver: University of British Columbia Press, 1995.

Mumme, Stephen P. "Dependence and Interdependence in Hazardous Waste Management Along the U.S.-Mexico Border." *Policy Studies Journal,* Vol. 14, No. 1 (September 1985), pp. 160–68.

Mumme, Stephen P., C. Richard Bath, and Valerie J. Assetto. "Political Development and Environmental Policy in Mexico." *Latin American Research Review,* Vol. 23, No. 1 (Winter 1988), pp. 7–33.

———. "State and Local Influence in Transboundary Environmental Policy Making Along the U.S.-Mexico Border: The Case of Air Quality Management." *Journal of Borderlands Studies,* Vol. 2, No. 1 (1986), pp. 1–16.

Northport Over Forty Club. *Northport Pioneers: Echoes of the Past from the Upper Columbia Country.* Colville, WA: Statesman-Examiner, 1981.

———. "One Hundred Years of Smelting at Trail." *Canadian Mining Journal*, Vol. 117, No. 5 (October 1996), pp. 10–17.

Oppenheimer, Michael, Charles B. Epstein, and Robert E. Yuhnke. "Acid Deposition, Smelter Emissions, and the Linearity Issue in the Western United States." *Science*, Vol. 229, No. 4716 (August 30, 1985), pp. 859–62.

———. *The People Who Will Live in Colville Area History.* Colville, WA: Statesman-Examiner, 1979.

———. *Phelps Dodge—A Copper Centennial, 1881–1981.* Special supplement in *Pay Dirt* (Summer 1981).

Power, Thomas Michael. *Lost Landscapes and Failed Economies: The Search for a Value of Place.* Washington, DC: Island Press, 1996.

Quinn, Mary-Lou. "Early Smelter Sites: A Neglected Chapter in the History and Geography of Acid Rain in the United States." *Atmospheric Environment*, Vol. 23, No. 6 (1989), pp. 1281–92.

———. "Industry and Environment in the Appalachian Copper Basin, 1890–1930. *Technology and Culture*, Vol. 34, No. 3 (July 1993), pp. 575–612.

Raftis, John T. *The Raftis Family of County Kilkenny.* Colville, WA, 1968.

Read, John E. "The Trail Smelter Dispute." *Canadian Yearbook of International Law*, 1 (1963), pp. 213–29.

Rice, Ross R. *Carl Hayden, Builder of the American West.* Lanham, MD: University Press of America, 1994.

Rieber, Michael. "The Economics of Copper Smelter Pollution Control: A Transnational Example." *Resources Policy* (June 1986), pp. 87–102.

Ritchie, Gordon. *Wrestling with the Elephant: The Inside Story of the Canada-U.S. Trade Wars.* Toronto, ONT: Macfarland, Walter, and Ross, 1997.

Roberts, T. M. "Long-Term Effects of Sulphur Dioxide on Crops: An Analysis of Dose-response Relations." In *The Ecological Effects of Deposits of Sulphur and Nitrogen Compounds*, ed. Sir James Beament et al., proceedings of a Royal Society discussion meeting, September 5–9, 1983. London: Royal Society, 1984.

Robinson, J. N. "The History of Sulfur Dioxide Emission Control at COMINCO Ltd." *International Journal of Sulfur Chemistry*, Part B, Vol. 7, No. 1 (1972), 51–56.

Rosenblum, Jonathan D. *Copper Crucible: How the Arizona Miners' Strike of 1983 Recast Labor-Management Relations in America.* Ithaca, NY: ILR Press, 1995.

Rubin, Alfred P. "Pollution by Analogy: The Trail Smelter Arbitration." *Oregon Law Review*, Vol. 50 (1971), pp. 259–82.

Saloutos, Theodore, and John D. Hicks. *Agricultural Discontent in the Middle West, 1900–1939.* Madison: University of Wisconsin Press, 1951.

Schmandt, Jurgen, Judith Clarkson, and Hilliard Roderick, eds. *Acid Rain and Friendly Neighbors: The Policy Dispute Between Canada and the United States.* Rev. ed. Durham, NC: Duke University Press, 1988.

Schwantes, Carlos A., ed. *Bisbee: Urban Outpost on the Frontier.* Tucson: University of Arizona Press, 1992.

Shea, Ira E. *The Grange Was My Life.* Fairfield, WA; 1983.

Shelton, Richard. *Going Back to Bisbee.* Tucson: University of Arizona Press, 1992.

Sheridan, Thomas E. *Arizona: A History.* Tucson: University of Arizona Press, 1995.

Smith, Toby. *Coal Town: The Life and Times of Dawson, New Mexico.* Santa Fe, NM: Ancient City Press, 1993.

Sorauer, Paul. *Handbuch der Pflanzenkrankheiten.* 3d ed. Translated by Frances Dorrence as *Handbook of Plant Diseases.* Wilkes-Barre, PA: Record Press, 1922.

Spelsberg, Gerd. *Rauch Plage: Zur Geschichte der Luftverschmutzung.* Aachen, Germany: Köner Volksblatt Verlag, 1988.

Stanford University. "Annual Report of the President of Stanford University for the Academic Year Ending August 31, 1925." *Stanford University Bulletin*, 5th series, no. 3 (January 1, 1926).

Stoklasa, Julius. *Die Beschädingungen der Vegetation durch Rauchgasses und Fabrikexhalationen*. Berlin: Urban and Schwarzenberg, 1923.

Swain, Robert E. "Atmospheric Pollution by Industrial Wastes." *Industrial and Engineering Chemistry*, Vol. 15, No. 3 (March 1923), pp. 296–301.

Swain, Robert E., and Arthur B. Johnson. "Effect of Sulfur Dioxide on Wheat Development." *Industrial and Engineering Chemistry*, Vol. 21, No. 1 (January 1926), pp. 42–47.

Thomas, Moyer D. "Effects of Air Pollution on Plants." World Health Organization monograph series no. 46, *Air Pollution*. Geneva: WHO, 1961.

Truett, Samuel. "Neighbors by Nature: Rethinking Region, Nation, and Environmental History in the U.S.-Mexico Borderlands." *Environmental History*, Vol. 2, No. 2 (April 1997), pp. 160–78.

Turnbull, Elsie G. *Topping's Trail: The First Years of a Now Famous Smelter City*. Vancouver, BC: Mitchell Press, 1964.

———. *Trail, 1901–1961*. Published by the Trail Diamond Jubilee Committee. Trail, BC: Hall Printing, [1961].

———. *Trail Between Two Wars: The Story of a Smelter City*. Vancouver, BC: Morriss Printing, 1980.

VanEenwyk, J. "Approaches to Community Concerns: Applied Public Health." *Public Health*, Vol. 111 (1997), pp. 405–10.

Wildavsky, Aaron. *But Is It True? A Citizen's Guide to Environmental Health and Safety Issues*. Cambridge, MA: Harvard University Press, 1995.

Wilkins, Mira. *The Emergence of Multinational Enterprise: American Business Abroad from the Colonial Era to 1914*. Cambridge, MA: Harvard University Press, 1970.

Wirth, John D. "The Trail Smelter Dispute: Canadians and Americans Confront Transboundary Pollution, 1927–41." *Environmental History*, Vol. 1, No. 2 (April 1996), pp. 34–51.

Yuhnke, Robert E., and Michael Oppenheimer. "Safeguarding Acid-Sensitive Waters in the Intermountain West: A Sulfur Pollution Strategy for Preventing Acid Pollution Damage in the Intermountain Air Shed." Boulder, CO: EDF, 1984.

INTERVIEWS

Babbitt, Bruce. Former governor of Arizona, 1979–1986. Tucson, AZ, June 9, 1988.

Bowerman, Lawrence. Known as "Mr. Smelter," he worked on smelters in the EPA's Air Quality Division from 1972–1985 in Region 9. San Francisco, CA, March 22, 1995.

Howekamp, David. Chief, Air and Toxics Division at EPA's Region 9. San Francisco, CA, November 10, 1994, and February 25, 1998.

Kamp, Richard. Director, Border Ecology Project. Bisbee, AZ, February 22, 1988 (with Michael Gregory in the afternoon), and February 23, 1988.

Leaden, Dorothy. Daughter-in-law of John Leaden, leader of the Citizens' Protective Association. Northport, WA, at the Leaden family farm, at Velvet Station, April 11, 1995.

Metzger, Cliff. State Department official who helped negotiate the La Paz agreement with Alberto Székely; later worked in the Office of Environment and Health. San Diego, CA, April 7, 1995.

Rieber, Michael. Professor of economics, University of Arizona. Tucson, AZ, June 9, 1988, and November 16, 1997.

Risner, William J. Tucson attorney active in pollution and smelter suits in the 1970s. Tucson, AZ, April 1, 1988.

Robinson, Priscilla. Founding director of Southwest Environmental Service, 1974–1988. Tucson, AZ, June 6, 1988, and November 16, 1997.

Scanlon, Matthew P. Retired Phelps Dodge Corporation vice president. Phoenix, AZ, November 18, 1997.

Székely, Alberto, and Roberta Lajous. As Consultor Jurídico and head of the North American desk, respectively, both were involved in various aspects of the La Paz agreement. Mexico City, DF, December 7, 1989.

Wick, Robert. Newspaper publisher. Telephone conversation with the author, October 29, 1997, and Sierra Vista, AZ, November 15, 1997.

Wrona, Nancy. Member of Governor Babbitt's staff who was responsible for environmental issues from 1981–1986; later worked at the Arizona Department of Health Services. Phoenix, AZ, April 1, 1988.

Yuhnke, Robert. EDF legal affairs specialist involved in western acid rain issues and the Douglas smelter. Boulder, CO, November 13, 1987.

INDEX

253